Energy for the
Year 2000

ETTORE MAJORANA INTERNATIONAL SCIENCE SERIES

Series Editor:
Antonino Zichichi
European Physical Society
Geneva, Switzerland

(PHYSICAL SCIENCES)

Energy for the Year 2000

Edited by

Richard Wilson

Harvard University
Cambridge, Massachusetts

Plenum Press · New York and London

Library of Congress Cataloging in Publication Data

International School of Energetics, 3d, Erice, Italy, 1979.
 Energy for the year 2000.

 (Ettore Majorana international science series: Physical sciences; 6)
 "Proceedings of the Third International School of Energetics, held at the Ettore
Majorana Center for Scientific Culture, Erice, Sicily Italy, September 1—11, 1979."
 Includes index.
 1. Power resources—Congresses. I. Wilson, Richard, 1926-   II. Title. III. Series.
TJ163.15.I567 1979   333.79   80-18623
ISBN 0-306-40540-7

Proceedings of the Third International School of Energetics, held at the Ettore
Majorana Center for Scientific Culture, Erice, Sicily, Italy, September 1—11, 1979.

© 1980 Plenum Press, New York
A Division of Plenum Publishing Corporation
227 West 17th Street, New York, N.Y. 10011

Printed in the United States of America

PREFACE

The Third International School on Energetics was devoted to
the subject of Energy for the Year 2000. By this title we hoped
to avoid discussion of such matters as the role of OPEC in raising
oil prices. In one sense, therefore, our task was made easier; we
could merely look into our crystal balls.

The choice of lecturers was made with the idea that no reasonable source of energy can be overlooked. We omitted detailed
lectures on oil and natural gas because we took it as a given fact
that we would continue to use as much of these fuels as we can get
at a reasonable price.

To give us an overview we started the School by discussing U.S.
energy policy and possible U.S. energy scenarios. As might be expected, there was some disagreement about the current energy program
in the U.S., but little disagreement about the facts presented.

Various energy options were examined. They included nuclear
power and the breeder reactor where the lecturers focused primarily
on controlling various problems inherent in using fission power for
energy production. There were lectures on coal--a source which will,
hopefully, be used in more and more environmentally superior ways,
and two sets of talks on solar energy--one a general survey and the
other a detailed discussion of photovoltaics. The talks on solar
energy were realistic about the current state-of-the-art and very
optimistic for long-term research and application. Energy production
using waves was a topic scheduled but was not presented due to illness.
However, the lectures do appear in this volume as originally prepared.

Richard Wilson

v

CONTENTS

ENERGY FUTURES: STRATEGIES FOR THE U.S.A.

Chauncey Starr
Electric Power Research Institute
Palo Alto, CA   94304

THE STRATEGY ISSUES

Introduction

The recent history of U.S. energy policy does not reveal a consistent strategy behind our actions. There are many reasons for this, but outstanding is the fact that no clear popular view has emerged to support a comprehensive strategy for dealing with the mixed aspects of the U.S. energy problem.

We have instead dealt with the discreet issues on which a consensus appeared, such as a 55-mile-per-hour speed limit, mandatory auto efficiency standards, and new building standards. Similarly, long-term research and development programs have been initiated, particularly in solar energy. Policy in both these cases could go forward politically before settlement of the difficult issues of oil pricing, environmental standards for coal mining and use, or any of the many controversial aspects of nuclear power.

This piecemeal approach and preoccupation with the detail of short-term factors has failed to assure the long-term availability

*Dr. Starr used materials from three reports in his lectures. The first is printed here, but due to copyright laws, we are unable to reproduce the other two. They were entitled, "Energy and Society" and "Energy Systems Options," and are contained in Current Issues In Energy, Chauncey Starr, Pergamon Press, New York, 1979.

of energy in the U.S. The means by which this objective should be
reached continues to be the center of the U.S. energy policy debate.

The purpose of this talk is, first, to consider the ways in
which the elements of the energy problem are generally understood,
to question that understanding, and to pose alternative statements
of the problem; and, second, to summarize specific aspects of energy
and electricity supply and demand which relate to strategy alternatives.

Conventional Energy Analysis

Conventional analysis of energy issues presumes generally that
economic cost-benefit analysis of the issues can produce a reasonable guide to optimum energy policies. Unfortunately, economic
analysis appears to be limited to the effects of changes that are
quite small relative to basic reference parameters, such as the
effect of small changes in energy price or supply. Effects of this
nature are referred to as marginal. The limitation of this type of
analysis is that large nonmarginal changes are obviously more important, yet conventional extrapolations generally result in the
misestimation of their effects.

In addition to the limitation that changes be small or "marginal,"
conventional analysis is also limited to those aspects of the problem
that can be quantified. For example, the point that economists
underestimate the social cost of energy supply has been a central
tenet of environmental groups, and they are correct. But there is
a corollary missed by the environmentalists--we similarly underestimate the social value of energy use. Neither has been well
quantified. By looking only at marginal changes, we equate the
benefits with the social value of the last unit of energy used, and,
of course, this generally tends to be wasteful or frivolous, as well
as the most costly unit.

Conventional energy analysis also frequently suffers from a
static view of social values. Those values which we currently hold
may change rapidly, as, for example, on the importance of environmental protection versus economic growth, or on the equitable control
of resources versus unrestricted competitive opportunity. As a more
familiar example, conspicuous, energy-consuming, large automobiles
have suddenly been displaced as status symbols by the diesel Mercedes
and the solar collector.

Complexity of Energy Issues

The mix of complex issues that goes into the formation of U.S.
energy strategy includes the following:

Supply Issues

o   The Geology and Geography of Energy Resources

o   Supply Technologies and Alternative Sources

o   Environmental Impacts and Public Risk

o   Imported Fuel--Cost, Availability, and Supply
    Security

Demand Issues

o   Conservation--Price Effects and Technology

o   Energy Needs of the U.S. Economy

o   Lifestyle Benefits of Energy Use

Political and Economic Issues

o   Impacts of Energy Imports

o   Equity and Distributional Effects vs. Aggressive
    Economic Growth

o   Energy Producers--Size, Power, and Profit

o   Demand Modification vs. Supply Modification

o   Energy Policy in Conflict with Social Goals

o   Energy Policy as a Surrogate for Social Policy

The length of the above list is a big part of the problem,
because the array of feasible strategies does not contain any
capable of resolving all these issues. Save for some possibility
that a lucky long shot will occur (for example, a miraculous photovoltaic discovery or extremely large domestic oil and gas finds),
no easy comprehensive solutions are visible. And those solutions
which require personal sacrifices (President Carter's "moral equivalent of war") have not been judged politically salable to the
American public--at least by the politicians.

Free Markets vs. Political Tension

Another aspect of conventional analysis is its assumption about
how world energy markets work. It treats energy trade (principally
oil trade) as if a free market exists. The attraction of this view
is obvious. The economic description of such a market is most

familiar to economists, and this model is a powerful analytical tool.

The free market view is clearly oversimplified and exaggerated. Certainly, even the strongest proponents of a free market recognize the existence of OPEC and the potential for using oil exports to influence political objectives. But what distinguishes the conventional view is an assumption that political decisions will bend to market forces, that is, that prices cannot be maintained at "artificially" high levels indefinitely.

The contrasting view is that we have a political world market, and that optimization from the seller's viewpoint does not mean maximizing only the present value of revenues, but includes influencing the foreign policy of the buyer. Lest we believe that political intervention of this sort is a recent or temporary effect, it is useful to recall that the U.S. has acted this way for a number of years in specific areas, for example, by trade embargoes with Cuba or by linking trade conditions with human rights policies. The point of these examples is not to critique such actions; it is rather common and that these policies have, in many cases, persisted for decades. The key issue for the U.S. raised by the political nature of the international market for fuels is the uncertainty of supply continuity, and the national security and economic vulnerability to a sudden and large reduction in supply.

Marginal Analysis vs. Supply Vulnerability

As an alternative to a policy based upon the premises of economic marginal optimization, it may be more important to the national interest to seek a policy which reduces both the likelihood and the effect of energy supply interruptions.

To some extent this position is exemplified by the decision made by several nations to create a strategic oil inventory. The U.S. has initiated such a reserve, recognizing that strategies should reflect the flexibility needed to deal with the political uncertainties.

But we failed to carry this understanding over to internal aspects of energy policy, despite painful evidence of the interruptible character of virtually all primary energy sources. In the past several years, shortages or interruptions of coal, natural gas, hydro, and nuclear power have occurred in addition to the oil embargo. It would be justifiable to subsidize high cost courses (and conservation) more aggressively to reduce the national vulnerability arising from dependence on any single source, either domestic or foreign.

On this basis, the U.S. would be wise to subsidize synthetic fuels from coal, oil shale, and liquid biomass fuels. If, as is occasionally alleged by proponents of these technologies, the lack

of private industry commitment to these systems stems from the
possibility that foreign oil producers can undercut their price,
then Government support guarantees are appropriate. It may well
be to the self-interest of the U.S. to pay $25/barrel for synthetic
fuels rather than $20/barrel to OPEC, when the intangible costs of
vulnerability to supply interruption are included.

We undertook our strategic oil reserve precisely because of the
insurance it provides against interruption, yet the issue of electric
utility reserve margin is rarely given the same consideration. In
the past, the rule of thumb that developed was that a 20 percent
reserve margin was roughly the desirable level to handle plant outages, maintenance, and uncertainty in demand due largely to the effect of extremes of weather. Today we have an added factor, supply
uncertainty. Since the 1973 embargo, the United States has experi-
enced several rather extreme weather conditions: the Western drought,
which reduced hydroelectric availability, and Midwestern blizzards,
which left coal barges stranded and coal stockpiles frozen solid.
In both cases the high reserve paid off, as the less affected oil
and nuclear capacity met the demand. The 1977 coal strike was another
case of supply interruption, as is the current mandated shutdown of
five nuclear plants for seismic analysis.

If we allow the implications of our intuitive understanding of
supply vulnerability, we should consider policies to reduce this
vulnerability by replacing insecure sources, such as imports, with
more secure domestic sources, even at higher prices. The obstacles
to such policies arise from several sources. First, and most apparent,
is the ideologic and political opposition within the U.S. to policies
which would accelerate either the rise in energy price or the expansion of coal and nuclear power, or early commitment (particularly with
Government guarantees) to synthetic fuels from coal. This aspect, like
the political component of the world oil market, is not quantifiable,
but is quite clearly a primary obstacle to reductions of imported oil.

A second aspect of expanding domestic sources to reduce imports
is the magnitude of the capital and other resource requirements needed
to do so. A rough estimate is that to replace all oil imports with
conservation, coal, nuclear, solar, oil shale, and other sources would
require an investment of about $250-$500 billion. Currently, direct
investments in the U.S. energy system are roughly 2 1/2 percent of
GNP, or $50 billion per year, and conservation investments add to this
unknown additional amount. But these conventional investment rates
have merely sufficed to meet some demand increases and compensate for
the depreciation of existing equipment. The investments of the past
several years have not even been sufficient to hold oil imports constant.

Attempting to achieve this substitution by the end of the century
would require that we double our investments into energy systems during

the next decade. This would represent a national peacetime investment less than half the military budget. President Carter's "moral
equivalent of war" description is, indeed, accurate in describing
the level of effort needed, if reduction of oil imports is the objective.

DEMAND AND SUPPLY PROJECTIONS

     At present we do not have an extraordinary U.S. effort to
accelerate alternative domestic energy sources which are technically
feasible and almost economically affordable (even if not yet competitive). In the absence of such a program, our energy demand and
supply projections continue to be based upon the criteria of commercial availability. Within the present conventional framework, the
projections for the U.S. here presented assume the acceptability of
oil imports and of expanding coal and nuclear power.

Demand Projections

     Historically, the growth of our economy and energy have been
closely coupled in both time and magnitude, as shown in Figure 1.
In the future, the relationship between these factors can be expected to be modified to some degree by conservation, new technology, and changes in the relative size of the service versus the
industrial sectors of our economy. A slow historical trend of reducing energy demand per unit of economic output is inherent in the
relationship displayed in Figure 1. This was primarily motivated

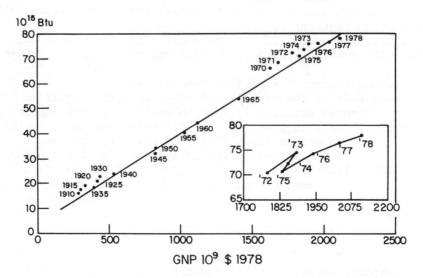

Fig. 1 Historical growth of energy use vs. economy

by the saving in capital cost of energy-converting and -using equipment which resulted from increased thermodynamic efficiency. The
recent increases in primary fuel costs have motivated supplementary
conservation efforts in addition to those in the historical trend,
and in this paper, the term "conservation" refers to this supplement.

    In recent years this strong relationship between energy and
economic output (as measured by GNP) has continued despite the 1973
oil embargo and the recession of the mid-1970's. As Figure 2 illustrates, GNP and energy have been tightly linked.

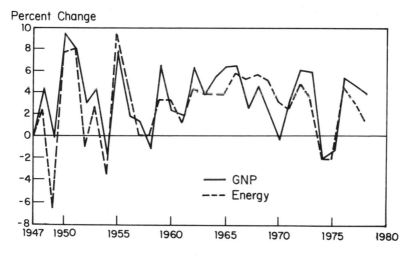

Figure 2.   Changes in energy and GNP.

We expect the coupling of energy and economic growth to continue
into the future, with the coupling coefficient depending on the
course of the principal variables. Four important variables to consider in terms of future energy requirements are: the productivity
of labor, employment level, the impact of conservation, and the
energy required to meet national air and water quality goals. We
know that the year 2000 labor force will be about 1 1/3 times the
present, because most are already born. In this analysis we have
assumed a 4 percent unemployment and a continuing trend of female
participation.

    Figure 3 shows a base case total energy requirement of 157
quads in the year 2000, assuming a plausible projection of economic
data and continuation of the historical relation to energy use.
The trapezoidal box shows the range of projections which occur
if the growth rate in the productivity of labor is varied between
0 - 2.3 percent per year and conservation is varied between 28-46

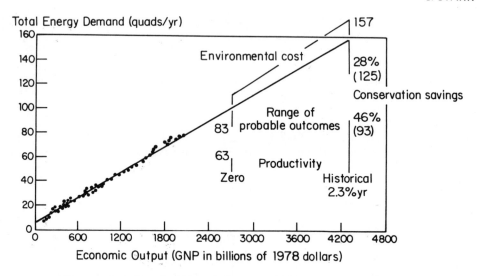

Figure 3.   Projection of year 2000 energy and GNP.

percent savings from the year 2000 base case. It is expected that
the pressure for environmental improvement will remain, and a 10
percent primary energy cost is estimated for this purpose.

There is a reasonable possibility that actual energy demand
and GNP will fall near the top of this box. The lowest demand
shown in the lower left-hand corner of the box indicates a year
2000 energy demand of about 63 quads, if 46 percent energy conservation could be achieved, and the productivity of labor were frozen
at today's value, a combination that is possible but unlikely.
Nevertheless, it is analytically correct that full employment could
be maintained without increasing the present levels, if individual
economic output is held fixed and conservation is pushed to its
technical limit.

Based on the importance of modest planning for a surplus rather
than a deficit, the upper right-hand corner of the box, which is
equal to 125 quads, may be the prudent target. This assumes 28 percent conservation by the year 2000, 10 percent environmental clean-
up cost, and historical growth in the productivity of labor. The
28 percent conservation savings appears to be an optimistic, yet
achieveable, objective. It should be realized that, as a nation,
we have been for many decades increasing the efficiency with which
we use energy, and that conservation savings much in excess of 28
percent will require either massive economic investment or significant technical changes. Economic pressures alone may motivate
the 28 percent conservation, as the cost of all primary fuels is
expected to increase steadily.

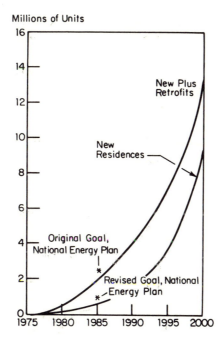

Figure 4. Solar heating of residences
(installation rate).

Without going into a detailed discussion of how this future
supply will be provided, it is valuable to examine the likely role
for technologies currently under development, but not yet commonly
in use. Of these sources, solar space and water heating is the
technology furthest advanced. Space heating with solar is somewhat
more expensive. An EPRI solar heating study analyzed the likely
energy savings under several different rates of utilization. This
is shown in Figure 4. As this figure indicates, an optimistic estimate is that 13 - 14 million residences would use solar heating
in the year 2000, these out of about 106 million residences projected.

The conventional energy displaced by this solar heating is
shown in Figure 5. As this study indicates, the fuel savings
amount to about one quad. Thus, even if we installed solar heating in 60 or 70 million residences, the savings would be about
5 quads. This could make a significant impact on natural gas consumption, but it only one part of the answer to our supply problems.

In considering how fast new energy sources can contribute to
supply, it is instructive to consider nuclear power as an example

for rapid integration. This is valid because, despite its current
problems, nuclear power had virtually unanimous support in its
early years. In those years plants were built much more rapidly
than today, because of the less detailed regulatory requirements.
Figure 6 indicates the path taken by nuclear power from the proof
of the scientific concept to its integration into utility systems.
Also illustrated in this figure are the paths for developing electricity options, assuming that integration will follow the nuclear
example.

Electricity Demand

    Perhaps the most difficult part of the job of supplying energy
in the amounts projected is the portion consumed as electricity.
This is because electricity consumption has been growing much faster
than total energy consumption, due to extremely long lead times
needed to build power plants and due to public opposition to the
expansion of any of our current major electricity sources: coal,
nuclear, and oil.

    The range of credible estimates for electricity demand growth
(taking into account both very extensive conservation and deeply
reduced economic growth) is roughly 3.8 - 5.5 percent per year,
which results in more than a doubling to tripling of annual demand
between now and the year 2000. Our pre-1973 experience in electricity growth was about 7 percent per year, which would have
quadrupled present production for the year 2000 electricity. Last
year's growth was 5 percent (April 1978 to April 1979). It may now
be lower, due to reduced economic growth. We should note that during
the national reduction in total energy use that occurred between
1973 and 1975, electricity growth increased. Electricity's uses to
the consumer are too great to keep the growth rate down.

    Even the minimum forecast (3.8 percent) leads to at least a
more than doubling of electricity generation (2.3 times) by the
turn of the century. How can a doubling of output be realized?
Obviously, the bulk must come from coal and nuclear.

    Coal plants now supply about 41 percent of our electricity
needs and consume 480 million tons of coal annually in the process,
two-thirds of a total production of 720 million. Considering all
the constraints of environmental regulations on end use, and the
delays and institutional constraints on increasing coal supplies,
by 2000 coal-produced electricity may be realistically limited to
slightly more than double that of today (a 5 percent per year growth),
perhaps providing a 45 percent share.

    The increasing difficulty of building coal power plants is not
generally understood. From time of decision to availability now

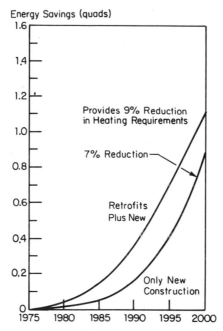

Figure 5.   Solar heating of residences
(energy savings).

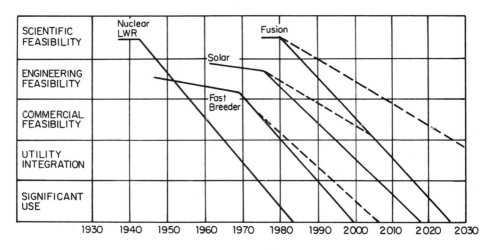

Figure 6.   Development phases for future power options.

takes eight to ten years, of which about half is used for the approval chain. Coal use also faces the Clean Air Act, the Resource Conservation and Recovery Act, the Toxic Substances Control Act, and the Clean Water Act. For new coal stations designed to meet the environmental requirements of those Acts, the estimated cost of environmental controls is about 60 percent of the total, and, as a result, the total cost of a coal station is now about the same as that of nuclear. Coal fuel costs are, of course, much higher than nuclear. Coal expansion may be further constrained, if it is extensively used to make synthetic liquid fuel. One ton may hopefully convert to three barrels of oil. So half of our present coal production converted to oil would produce about one billion barrels of oil per year--about 16 percent of our present use and about a third of our imports.

Hydroelectricity provides about 11 percent of our generation output now, and could possibly be increased during the next twenty years, but not to the extent of doubling, so that it may provide about 7 percent of the output in 2000. Geothermal supplies are about 0.2 percent now, and hopefully will be about 2 percent in 2000.

With regard to the solar distribution, the February 1979 report to the President of the Domestic Policy Review of Solar Energy presents an interagency forecast of the year 2000 contribution of solar electricity (thermal, photovoltaic, and wind) based on very optimistic (in some cases unrealistic) assumptions of successful technical development. Their projection for an equivalent energy displacement by solar is 2 - 6 percent of the minimum year 2000 electricity demand we project.

The total from these sources is 56 - 60 percent of that required to meet the low projection for the year 2000. The remainder will need to come from nuclear, synthetics, oil, and gas. At present, about a third of our generation depends on oil and gas. Given our national need for liquid fuels for transportation and the strong federal policy to diminish their use for electricity generation, it is unlikely that synthetics and new oil and gas can be considered for electricity expansion purposes. Oil and gas will probably generate somewhat less electricity than today, perhaps 13 percent of the year 2000's minimal needs. We are left with about a fourth of our needs to be provided, even with our minimum growth estimate, and nuclear is the only source that can fill this gap.

The year 2000 shortage, if nuclear is not pursued, is substantial. This is more easily perceived in terms of the number of nuclear power plants involved. A nominal plant would be about 1 GWe (Gigawatt) size (one million kW), and costs in the neighborhood of one billion present dollars. The equivalent generating capacity in the year 2000 to meet our minimum projection is about 1,200 GWe, so the gap we need to fill by nuclear is roughly 300 GWe. Nuclear

stations now operating and soon approaching completion (if that is
permitted) will supply only about 100 GWe, so the shortage will be
about 200 GWe on the lowest growth assumptions for electricity demand. Note that this is about 10 years of a reasonable and maximum
annual construction target. Because of regional variations, the
impact on the high-growth areas of the country would be much larger.
If the most recent rate of growth of 5 percent were to continue,
the shortage would be about 600 GWe. Recognizing that these forecasts already include more than a doubling of coal generation, you
can understand why utilities perceive nuclear not as a matter of
preference, but rather, as a crucial necessity.

One point should be understood concerning the utility view of
solar energy. As these estimates indicate, the utilities believe
they will need every bit of generation capacity that they can find,
and they therefore have no reason to downplay the role of solar
power. In fact, when solar-electric generation arrives, it will
probably be used for the intermediate portion of the load currently
filled by oil. But it is not yet available at costs that are even
remotely acceptable to customers--present costs are 10 to 20 times
conventional delivered costs. Only wind supplement is near to
approaching competition with oil. The idea that solar-electric
could be a pre-2000 source of base-load power, capable of eliminating the need for either nuclear power or of expanded coal use,
is totally false.

The shortfall without nuclear energy is at least 25 percent
by year 2000, and may range up to twice that at the higher growth

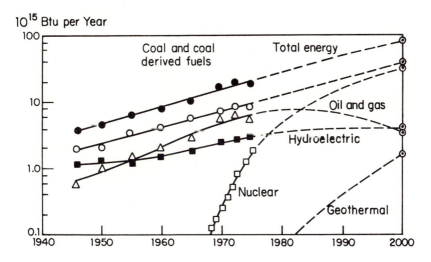

Figure 7.   Energy input for electricity (Year 2000 planning targets.)

rate. This gap may take the form of measurable physical shortages, or it may be partially absorbed by the gradual adjustments to constrained supply that will occur as a chronic condition in the 1980's and 1990's. Conversely, if the use of both coal and nuclear power is expanded, the demand can be met. Figure 7 illustrates the mix of sources that could plausibly be used.

Conclusion

Given all these uncertainties in the framework for U.S. planning--both as to the resource assumptions and as to the objectives and their priorities--what should be the U.S. posture? A pragmatic policy, free of ideologic content, would be to seek implementation of a large diversity of energy sources. Those that are now commercial are clearly the most certain and should be fully deployed. Those that are experimental should certaintly be brought to the demonstration stage. The vulnerability of a nation to an energy shortage is so large that the economic costs of energy research, development, and commercialization are relatively modest by comparison. It is a national insurance policy which is both affordable and necessary.

Unfortunately, the U.S. energy debates have politicized the development of energy alternatives--renewable versus non-renewable sources; solar versus the breeder; centralized versus decentralized; conventional versus exotic, and so forth. A propagandized and confused public can hardly separate reality from fantasy, present availability from future hopes. When political leaders simultaneously hinder oil development, constrain coal use, and limit nuclear, one wonders which world they live in. Further, surprisingly for a nation whose history enshrines private enterprise, government control of the energy sector in the past few years has become overpowering--so much so that market competition among energy sources has become highly distorted, and conventional evaluations of alternatives have lost their significance.

Facedwith the probability of such a dismal outcome resulting from our present national posture, what should we be doing now to avoid it?

I believe that the nation must develop new energy concepts for future deployment beyond 2000, and expand every existing energy and electricity source that technology has brought within range of economic feasibility. A "must" is extensive deployment of nuclear power, which still has a U.S. industrial infrastructure sufficient to close the electricity gap that would otherwise occur although it is slowly fading away. Recognizing the uncertainties of future alternatives and the on-shot nature of the conservation opportunities, we cannot, in common sense, neglect the practical option of nuclear power.

SOLAR PHOTOVOLTAIC ENERGY CONVERSION

H. Ehrenreich

Division of Applied Sciences
Harvard University
Cambridge, MA, USA   02138

GENERAL INTRODUCTION

The lectures presented under this title consisted largely of
material drawn from the American Physical Society's Study Group on
Solar Photovoltaic Energy Conversion. The panel was chaired by the
author. The other members were D. DeWitt (IBM), J.P. Gollub
(Haverford College), R.N. Hall (General Electric), C.H. Henry (Bell
Labs), J.J. Hopfield (Princeton University), T.C. McGill (CalTech),
A. Rose (Boston University), J. Tauc (Brown University), R.M. Thomson
(National Bureau of Standards), M.S. Wrighton (MIT), and J.H. Martin
(Harvard University), who served as Executive and Technical Assis-
tant to the Study. The following material is abstracted from the
Study Group's report on which the lectures focused, with some
editorial alterations from the original. The reader should refer to
the published report, "Solar Photovoltaic Energy Conversion" (1979,
American Physical Society, 35 E. 45th St., New York, NY 10017) for
definitive information concerning the Study Group's conclusions.
The following precis, however, should serve to convey the substance
of these conclusions.

The lectures also included background material concerning
photovoltaics, but this is widely treated in the literature. There
is a list of some general references at the start of the reference
section.

The material is organized around five summary statements, which
may be regarded as extended chapter headings. In the APS report
these statements formed part of the executive summary.

1.   It is unlikely that photovoltaics will contribute more
     than about 1% of the U.S. electrical energy produced near
     the end of the century. Central power production is the
     most extensively studied and clearly perceived long-term,
     large-scale application of PV for this country. Barring
     unforeseen rises in the cost or availability of fuels,
     prices for 12-16 % efficient flat plate modules or
     concentrator arrays of about 10-40¢ per peak watt (W_p)
     in 1975 dollars will be required to compete with the
     projected cost of coal-generated electricity (about 45-70
     mills/kWh levelized busbar cost in the year 2000).

INTRODUCTION

Because of the uncertainties in the future cost of the alternatives for electricity generation and in the future cost of PV
systems, the extent of the photovoltaic contribution to future U.S.
electrical power generation is not known. The probable range of
competitive PV costs during the next few decades will be discussed
in this section. These estimates are based on a synthesis of information contained in the major systems design studies that have been
performed to date and various cost estimates for the fabrication of
solar cell modules.

It will be seen (Table 3) that the systems design studies are
mutually consistent within a factor of two. We estimate forecasts
for silicon module production costs to have an associated uncertainty of perhaps −20 to +100 %. It is to be emphasized that in all
cases these estimates represent projections of the present technology which include technological improvements of various kinds and
increased automation, but not breakthroughs.

The year 2000 has been commonly chosen as a time period in
which central power PV in the U.S. might first become significant,
because it is about a decade after the development of present
technology would have reached fruition. Within this scenario, the
last decade of the century or so would remain for large scale production of PV systems. It seems reasonable to call a power source
"significant" if it provides 1 % or more of U.S. electricity. Table
1 presents a rough estimate of the requirements for PV to provide
that amount of generation in the year 2000.

The results shown in Figures 1 and 2 summarize the conclusions
of this section. The captions give self-contained descriptions of
these figures. Except for the cost of coal electric generation, the
uncertainties resulting from the absence of accurate cost infor-

mation are not indicated. They should however be kept in mind by the reader, as should the fact that the conclusions derived from the input information represent the Study Group's judgment in all cases.

ALLOWABLE COST RANGE FOR PV SYSTEMS

A build-up at uniform rate to the 1 % level in ten years would require a PV production rate of 2000 MW_p per year beginning in 1990, more than 1000 times our present level of production. Because of the time required for public and utility acceptance, production scale-up, development of automated factories, and most importantly inventions needed to reduce costs, we do not expect the 1 % level to be reached much before the year 2000 unless an emergency deployment of PV is necessary. We will therefore try to estimate the costs of conventional and PV generation in the 2000-2030 time period.

Barring unforeseen developments, the large-scale use of PV in the U.S. will not occur unless PV electricity is competitive in cost with conventional forms of electric power generation. Only then will PV be used by utilities to supply electricity to the power grid or be purchased in significant quantities by customers who have the option of receiving power from the grid.

The market potential for nongrid-connected (remote) applications has been studied by the Solar Energy Research Institute (SERI).[2] While this market can be served by higher-cost PV systems, it is quite uncertain in magnitude (30-340 MW_p/year) and rather small in comparison to U.S. electricity consumption. Furthermore, much of this market comprises applications, such as pumping water in foreign countries, which provide no direct fuel savings to the U.S.

Several major conceptual design studies of grid-connected PV systems have been made for Sandia,[3-6] for JPL,[7] and for the Electric Power Research Institute (EPRI).[8] Grid-connected applications can be roughly divided into central power station and on-site applications, the latter including the residential, commercial, and industrial sectors.

The central power application of PV is reasonably well defined at this time. There is rough agreement on the cost of the non-module parts of the system and the methodology[9] for evaluating the life cycle cost of PV and conventional systems. Thus, the allowed capital investment CI in cents per watt of peak capacity (c/W_p) for competitive PV central power systems and the allowed module costs C_W in c/W_p can be approximately calculated. In Section 2 we point out that PV can be expected to displace mainly intermediate load generation. Therefore, we will determine the range of allowed CI by comparison with the expected price range of future intermediate coal generation.

Table 1

Requirements for PV to contribute 1 % of U.S.
electricity generation in the year 2000

1975 U.S. electricity generation	$1.92 \cdot 10^{12}$ kWh
Plausible U.S. electricity generation in 2000[1] (3.9 % projected growth rate)[a]	$5 \cdot 10^{12}$ kWh
Significant PV generation (1 % of total generation)	$5 \cdot 10^{10}$ kWh
Required PV capacity in the Sunbelt[b] (1 W_p produces 2.5 kWh/year)	20 000 MW_p
Annual production rate for 10-year build-up	2000 MW_p/year
Thermal fuel displaced (1 $kW_{electric}$ = 2.9 $kW_{thermal}$; 1 Quad = 10^{15} BTU)	.50 Quad/year
Capital investment (50–100 ¢/W_p system cost, 1975 $)	$1–2 \cdot 10^9$ $/yr
Array area (10 % efficiency from light to electricity)	200 km^2
Total land area required	400–600 km^2

[a] Moderate growth scenario; most current estimates lie in the
range from 0 to 6 % per annum.

[b] The rating of a PV plant in peak watts (W_p) is the electrical output of the plant with an insolation of approximately
1 kW/m^2.

The on-site applications are less well defined at present. The installed cost per watt of these systems will probably be greater than for central power systems.[10] Although there are savings associated with the use of roofs as support structures in some designs, installation costs will be higher than for central stations and the price of system components at the "retail" level will be higher. The smaller scale of on-site systems results in higher relative costs of power conditioning. Either expensive storage and local back-up or back-up connection to the utility grid is required. Because of the erratic nature of residential loads, residences must throw away excess PV energy, store it, or sell it back to the utility. When a residence is grid-connected and returning excess power to the grid, it becomes in effect a small costly central power station.

Unless current development practice can be modified, the adoption of residential systems will also be hampered by the lack of suitable roof orientations, by the constraints this requirement places on the design of new homes powered in part by PV, by the difficulty of using tracking concentrators residentially, and by the burden of system maintenance on the home owner.

The major advantage of on-site applications over central power PV is the opportunity for cogeneration of heat and electricity using a single system. However, studies of cogeneration in residential systems[4,6,11] show no significant cost advantage of combined PV and thermal arrays over side-by-side generation of electricity and heat. Cogeneration does appear attractive for application at large load centers or for industrial sites where high concentration liquid cooled PV systems are employed. Preliminary studies[12] indicate a definite advantage when all the thermal heat generated can be used.

It is not yet established whether on-site PV with cogeneration is advantageous compared with central power in the U.S. Because of the existing uncertainties, we have used studies of central power systems to determine the range of competitive module costs.

In summary, we have studied the case of PV use in central power generation because the application requirements are quite well defined, the potential market is large, either flat plates or concentrators may be used, and the cost requirements for PV modules used in central power appear no more stringent than, for example, residential U.S. PV applications.

COST REQUIREMENTS FOR COMPETITIVE PV POWER

The levelized busbar energy cost $\overline{BBEC}$ is a figure of merit used by utilities to compare the economic value of alternative generating options. It is the price per unit energy which, if held constant

throughout the life of the system in current dollars, would provide
the revenue required to meet all costs of the system. We will use
the levelized busbar energy cost, expressed in 1975 mill/kWh, to
compare PV plants and intermediate-load coal plants operating in the
2000–2030 time period.

The $\overline{\text{BBEC}}$ of a generating station is given by[9]

$$\overline{\text{BBEC}} = \frac{10^4 \ \text{FCR} \cdot \text{CI}(\text{¢}/W_p)}{H_A(h)} + \overline{\text{OM}} + \overline{\text{FL}} \tag{1}$$

where $\overline{\text{BBEC}}$, $\overline{\text{OM}}$, and $\overline{\text{FL}}$ are measured in mills/kWh (1 mill = \$.001)
and we have defined

CI: the initial capital investment

FCR: the fixed charge rate, an annual percentage rate
covering taxes, interest and return on equity,
repayment of principal, and insurance

H_A: the effective number of hours per year of operation
at rated capacity (the annual plant output in kWh
divided by the plant's rated capacity)

$\overline{\text{OM}}$: the levelized cost of operations and maintenance

$\overline{\text{FL}}$: the levelized fuel cost.

Typically, FCR = 0.15/yr. For an intermediate load coal plant,
H_A is nominally 4380 h/yr (capacity factor 50 %).[13] With these
assumptions and a CI estimated by EPRI[13] as (67 ± 9) ¢/W_p for
a coal plant with SO_2 scrubbers, the first term in Eq. (1) is
(23 ± 3) mill/kWh. EPRI[14] has estimated $\overline{\text{OM}}$ = 4 mill/kWh.
Using these figures the range of $\overline{\text{BBEC}}$ for future coal plants is then
specified by the estimates of the future price of coal.

Table 2 shows a variety of coal price estimates and the resulting levelized busbar costs. The references in the table give the
sources of the coal estimates. The range of busbar costs extends
from 44 to 72 mill/kWh. To be competitive in this application, a PV
plant must have a value of $\overline{\text{BBEC}}$ which falls into this range unless a
large value is assumed for the so-called external costs of coal generation. A plant efficiency of 10^4 Btu/kW_eh, corresponding to a
thermal efficiency of 34 %, was used to calculate $\overline{\text{FL}}$. The range of
$\overline{\text{FL}}$ corresponds to a range of 2.0–4.2 \$/MBtu (1 MBtu = 10^6 Btu) for
the levelized price of coal (1975 dollars).

The cost of coal delivered to electric power plants in 1975 was 86 ¢/MBtu for the U.S. average and 56 ¢/MBtu for the Southwest. If this cost remained constant in real terms, rising only at the general inflation rate assumed, $\overline{FL}$ would be 16 mill/kWh (U.S. average) and 11 mill/kWh (Southwest) in the period considered. The higher values for $\overline{BBEC}$ and $\overline{FL}$ in Table 2 are associated with assumed increases in the real cost of coal up to and during the 2000-2030 time period.

In order to evaluate the allowed capital investment of a PV plant, we have calculated the $\overline{BBEC}$ of a PV central station plant using the same FCR (15 % per annum) as for coal plants, assuming a nominal insolation of H_A = 2500 h/yr, corresponding to the most favorable areas of the U.S. Southwest, and estimated operations and maintenance costs of 4 mill/kWh.[7,15] The same annual insolation was used for flat plate and 2-axis concentrator systems because, while the latter only utilize the direct component of the sun's radiation, they remain facing the sun throughout the day. These effects approximately cancel each other in most of the U.S.[12]

In comparing PV and conventional plant electricity costs, one must be careful to take into account that the PV plant must contribute the same amount of effective system capacity as the conventional plant it displaces. One way to do this is to add gas-turbine back-up generation during cloudy periods to insure that the PV plant is as reliable as a conventional plant. A 10 % PV energy penetration in the U.S. Southwest requires about 125 hours/year of back-up generation,[6] costing about 9 mill/kWh when averaged over the 2625 effective h/year of power generation at rated capacity. This approach leads to a break-even capital investment CI in the PV plant of 52-98 ¢/W_p, the range resulting from the spread in levelized BBEC of conventional coal plants indicated in Figs. 1 and 2.

Another approach we took to determining the allowed CI of PV plants was based on the results of a study[8] which concluded that in the Southwest the effective capacity of a PV plant with no back-up is about 50 % that of a conventional plant. Therefore 1 kW_p of PV capacity displaces about .5 kW_p of conventional plant. This approach leads to a CI = 50-93 ¢/W_p, in good agreement with our other estimate.

FORECAST OF PV SYSTEMS COSTS

A PV plant beginning operation in the year 2000 will make use of PV technology developed and in manufacture by the mid 1990's. There are no accurate cost estimates available for PV system components in that period. We have tried to estimate the minimum costs of some of today's most clearly defined PV options for the mid 1990's by making use of systems studies, of the detailed cost goals of the Si flat plate program, and of cost information on low cost heliostats designed for power tower applications.

Non-module Costs

The cost of a PV system can be divided into the module cost (cost at the factory of the flat plate panel or concentrator) and the cost of the remainder of the system. Table 3 gives two estimates of the components of the non-module system cost, along with the numbers used here. The first column is an average over several designs in a recent detailed engineering study for JPL.[7] The second column is a Sandia summary[19] of a number of recent studies. Our estimate of only 20 $/m^2 for land, structures, and installation is much less than the 38.5 $/m^2 of the engineering study,[7] which is based on current technology. This lower number does not derive from an engineering study. It is attainable only if low-cost structures and highly automated installation methods, such as those described in a recent study of an enclosed heliostat design,[18] are developed.

Table 2

Various projections of the cost of intermediate load
coal electric generation in 2000-2030
(1975 dollars)

	EPRI [14]	DOE (high) [15]	DOE (moderate) [15]	GE (high) [16]	GE (low) [16]	Aerospace Southwest US [17]
FL (mill/kWh)	20-26	41	21	35	27	28-42
BBEC (mill/kWh)	44-56	65-71	45-51	59-65	51-57	52-72

Table 3

Non-module system costs:
near-future estimates and cost model
for central power stations

Item	JPL-Bechtel[7]		Other Conceptual Design Studies[19]	Cost Model Used Here
Power Conditioning ($/kW) (1975 dollars)	AC wiring 1/2 DC wiring Converter Switch yard	5.6 5.6 66.0 66.0 —— 143.2	60–125	140 $/kW
Land, Structures, Installation ($/m²)	Land Clear and Grade 1/2 DC wiring Lightning protection Foundations Support Frame	0.8 4.3 0.8 1.6 9.6 4.3 17.1 —— 38.5	20–40[a]	20 $/m²
Indirect Costs (%)	Engineering Contingencies IDC[b] and other	5.6 8.2 30.6 —— 44.4	10 5 15 —— 30	35 %
Operation and Maintenance (mill/kWh)		3.3	5	4 mill/kWh
Plant Efficiency (%)		91	95	90 %
Back-up[c]				9 mill/kWh

---

[a] 10 % overall efficiency for converting sunlight to busbar AC electricity assumed

[b] Interest during construction

[c] Back-up generation is assumed to contribute 125 kWh/year per kW of capacity; the cost model figure is averaged over the total PV system electricity generated and includes both capital and fuel costs for the back-up.

Flat Plate System Costs

    The non-module costs of Table 4 are used to plot the capital
investment required for the PV system $CI(\text{¢}/W_p)$ as a function of
area related costs in Fig. 1 for flat plate systems. For such a
flat plate system

$$CI(\text{¢}/W_p) = .1 \, f_{ind} \, (\frac{C_S + C_A}{e_p \, e_m \, I_p} + C_{PC}) \quad , \quad (2)$$

where we have defined

                                                         cost model value

    f_{ind}: indirect cost factor (see Table 3)   1.35

    C_S: site cost (land, structure,
                installation, etc.)   20 $/m^2$

    C_A: area cost of modules in $/m^2$ variable

    e_p: plant efficiency for module output
                to busbar AC   .9

    e_m: module efficiency for transforming light to
                electricity at the module terminals   variable

    I_p: peak insolation   1 kW/m^2

    C_{PC}: power conditioning capital cost   140 $/kW_p$

The .1 is a conversion factor: $1 \, \$/kW_p = .1 \, \text{¢}/W_p$. The module
cost per peak watt, $C_W(\text{¢}/W_p)$, is related to the module area
cost, $C_A(\$/m^2)$, by

$$C_W = \frac{C_A}{10 \, e_m \, I_p} \quad . \quad (3)$$

Eliminating the module efficiency e_m from Eq. (2) using Eq. (3)
gives

$$CI = f_{ind} \, (\frac{C_W}{e_p} (\frac{C_S}{C_A} + 1) + .1 \, C_{PC}) \quad . \quad (4)$$

The curved dashed lines in Figs. 1 and 2 correspond to constant
module cost C_W. As the module efficiency is decreased, C_A
decreases for constant C_W and CI increases, as Equation (4) shows;
CI diverges as C_A approaches zero.

Table 4

Flat plate module costs:
1986 DOE goals and cost model for
central power stations

	JPL 1986 Technology Goals[20] 46 ¢/W_p 11.5 % Efficiency Crystal Si Flat Plate Module		Cost Model for Thin Film Flat Plates
Cell Cost[a]			
Metallization	7.0 ¢/W_p	8.0 $/$m^2$	6.5 $/$m^2$
A-R Coating	0.8	0.9	1.0
	8	9	7.5
Noncell Cost			
Interconnection	4.2	4.8	3.5
Encapsulation	12.0	13.8	14.0
Test	0.1	0.1	
Package	0.1	0.1	
	16 ¢/W_p	19 $/$m^2$	17.5 $/$m^2$

[a] The costs of silicon preparation, sheet fabrication, p+ back
layer formation, etching, ion implantation, and pulse annealing,
which form 22 ¢/W_p of the JPL total of 46 ¢/W_p, are omitted
since these processes will not carry over to thin film cells.

Table 5

Concentrator system costs:
cost model for central power stations

	Cost per Unit Area of Aperture
Non-Module Cost	
Land, clearing/grading, wiring, lightning prot.	12.5
Installation of heliostat[18]	7.5
	20.0 $/$m^2$
Module Cost	
Heliostat[18]	17.5
Additional cost for parabolic reflector shape	15.0
Cooling	10.0
	42.5 $/$m^2$
Cell Cost	
Si cells (250 $/$m^2$ of cells at conc. ratio C = 50)	5.0 $/$m^2$
GaAs-GaAlAs cells (7500 $/$m^2$ of cells at C = 500)	15.0
Multicolor cells (10/000 $/$m^2$ of cells at C = 500)	20.0

The diagonal shaded regions in Fig. 1 indicate the projections of the possible range of cost of future flat plate systems. The shading for Si flat plates ends at $C_W = 50¢/W_p$, the 1986 DOE cost goal for the Low-Cost Silicon Solar Array Project (LSSA) managed by JPL.

The cost goals of the 1986 Si flat plate program, excluding the cost of Si wafers and junction formation, are listed in the first two columns of Table 4. Similar costs will be encountered for thin film flat plates. As a result, the minimum area costs of a thin film module are estimated in the right-hand column of Table 4 to be only slightly less than the corresponding cost items taken from the cost goals of the Si program. We attribute an interconnection cost to the thin film flat plates, because while large area thin film cells may be fabricated, many small cells wired in series are needed in order to keep module currents reasonable.

The largest part of the module cost is 14 $/m^2 for encapsulation. This estimate is based on the assumption that hermetic glass encapsulation will be required for reliable thin film systems. If durable thin film modules encapsulated in plastic could be developed, this cost might be substantially reduced, but whether this can be accomplished is an open question.[21] Our minimum estimated cost for the thin-film module of 25 $/m^2 is obtained by adding the cell and noncell costs in Table 4 and assumes negligible cost for the thin film itself. This value determines the cut-off of the shaded thin film region in Fig. 1. Figure 1 shows that the nonfilm area related costs require minimum thin film module efficiencies of near 10 %.

Concentrator System Costs

The minimum costs for two-axis concentrators are estimated in Table 5. This table is based on two designs for low-cost enclosed heliostats[22,18] for power tower collectors, made of Mylar-like plastic sheet. Each design uses a plane reflector of about 60 m^2 area enclosed in an air-supported plastic enclosure. Zimmerman[22] has recently modified one of these heliostat designs for PV use, replacing the flat reflector with a lightweight parabolic plastic reflector which can focus sunlight on PV cells at high concentration. In Table 5 we list our own rough estimates of the additional cost expected for cooling the cells with a circulating fluid and for modifying the reflector into a focusing element.[23] The cost estimates used in Fig. 2 are only expected for large volume production, on the order of 250,000 concentrators (about 3000 MW$_p$) per year.

The installed concentrator cost, including metal foundation but without cells, is estimated to be 50 $/m^2 (42.5 concentrator + 7.5

installation; see Table 5). This is much less than current manufacturing costs of conventional concentrators: the current installed
cost of a two-axis tracking Fresnel lens concentrator is estimated to
be 246 $/m^2 without cells. For comparison, a 50 ¢/W_p 12 % effi-
cient flat plate module would cost 60 $/m^2. The enclosed plastic
concentrators have the potential for extremely low cost at high
volume because they are made of lightweight inexpensive materials.
While these plastic structures are extremely attractive because of
their low-cost potential, they have not yet been proven to be durable
under service conditions. Problems could arise, such as the degradation of the optical quality of the plastic enclosure after several
years. Their success is therefore not assured. However, the plastics industry is only now beginning to consider this application and
better enclosure materials may be developed. This area of materials
research is of great potential importance.

The expected CI for these concentrator systems was calculated
from Eq. (2) and is plotted in Fig. 2. We assume an optical concentrator efficiency of .7 (e_{module} = 0.7 e_{cell}) and cell operating
temperatures of 50 C. The cell efficiencies at 50 C are shown on the
figure. The lower efficiency in each shaded region corresponds approximately to the current state of the art. The higher efficiency
in each case is our estimate of what might be achieved. The cutoff
of each shaded region is determined by the cost estimates of Table 5.

No correction has been made in Figure 2 for the power requirements of the cooling system. Cooling of cells has been estimated[11]
to consume about 3-15 % of the output power, depending upon the
system constraints.

CONCLUSIONS

Figs. 1 and 2 show that the allowed cost of PV systems operating
in the U.S. Southwest (H_A about 2500 h/yr) will be 50-100 ¢/W_p.
The corresponding range for average-insolation areas of the United
States (H_A about 1750 h/yr) would be 35-70 ¢/W_p. This range is
largely determined by the uncertainties in future coal prices. From
Figure 1 it can be seen that the allowed factory price of modules
necessary to meet these system cost requirements is 10-40 ¢/W_p.

The relative potential of each PV option should be judged by the
required size of its capital investment. Reasonable lower limits are
shown by the cutoffs of the shaded regions in Figs. 1 and 2. These
cutoffs have been determined by considering each option using the
same degree of "reasonable optimism" as a criterion. However, in the
absence of accurate cost information these graphs represent our own
judgments in all cases.

According to our projections, Si flat plate modules will not be competitive at the 1986 DOE goal of 50¢/W_p module cost. This cost level will principally provide insurance against unexpectedly high future fuel costs of economic, environmental, or political origin.

Thin film flat plates have the potential to become competitive if modules with efficiencies near 10 % can be developed, if the cost of the thin film without contacts can be kept down to a few $/m^2, if low-cost support and installation costs are achievable, and if electricity prices are high.

Si cells used in inexpensive concentrators are promising candidates for competitive electricity generation. The main obstacles to the success of this approach are the development of low-cost two-axis concentrators and the production of Si concentrator cells with efficiencies near 20 % at a temperature of 50 C.

High efficiency cells, such as improved GaAs-GaAlAs cells or multicolor cells, appear to offer the greatest potential for becoming competitive at currently projected fuel prices, provided that extremely low-cost concentrators capable of operation at a concentration of about 500 can be developed and that adequate service life of cells and concentrators can be achieved.

Concentrators appear to have an advantage over flat plates. The flat plate approach requires deployment of large areas of low-cost and durable materials, which at the same time are sophisticated electronic devices capable of converting light efficiently into electricity. The concentrator approach separates the conversion problem into two parts. To make use of the highly collimated nature of direct sunlight, inexpensive durable materials are used to focus the sunlight on a small efficient PV converter. A very sophisticated converter (which is costly per unit area) can then be afforded.

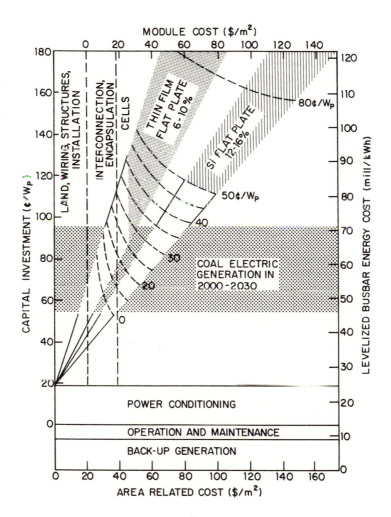

Figure 1

Allowed costs for flat plate PV systems used in
Southwest U.S. central power stations, circa 2000-2030

The vertical axis on the right-hand side of the figure corresponds to the levelized busbar energy cost. The horizontal shaded band represents the probable range of levelized busbar energy cost for future intermediate load coal plants. It thus determines the probable range of allowed capital investment for competitive PV systems of equivalent reliability.

The left-hand vertical scale gives the capital investment per peak watt in a PV system which produces a given busbar energy cost in the U.S. Southwest (see Eq. 2). The left and right scales do not have the same zero because of the operation costs required by the model PV system and the necessity for back-up generation to insure reliability equal to that of an equivalent coal plant. The back-up cost is indicated at the bottom of the graph.

Area-related costs are plotted on the horizontal axis. Non-module (land, wiring, structures, installation), noncell (interconnection and encapsulation), and cell costs per unit area are separated by vertical dashed lines. The diagonal lines represent the linear relation between the capital investment cost per peak watt of PV plants and the cost per unit area as determined by Eq. 2 for various module efficiencies. Efficiency ranges of 6 to 10 % for thin film modules and 12 to 16 % for silicon modules are shown.

The curved dashed lines plot the system capital investment for constant module cost per peak watt. The lines curve upward as module efficiency decreases because greater collector area is required for a given output and the area related non-module costs raise the cost of the total system. The vertical line corresponding to zero module cost represents the asymptote.

The shaded portions of the pie-shaped regions referring to thin film and Si flat plates respectively represent cost ranges that are attainable by straightforward extensions of present technology. The demarcation line between shaded and unshaded portions, corresponding to the lowest attainable cost, is better visualized as a smeared-out region reflecting a healthy degree of uncertainty resulting in part from the choices of the cost model given in Table 3. The limit for thin film cells is determined by assuming negligible cost for the film itself, the cell costs being those associated with metallization and antireflection coating as shown in Table 4.

For silicon the range extending to the 1986 DOE module cost goal of 50 $\cent/W_p$ has been shaded. It appears that major technological advances will be required to achieve significant reductions below 50$\cent/W_p$.

As an example of how to interpret Fig. 1, consider a thin film module with an efficiency of 10 %. Our projected lowest costs for this system are given by the lowest point on the right-hand side of the shaded region for thin film flat plates: a module cost of 25 ¢/W_p (the curved dashed line) or 25 $/$m^2$ (from the upper horizontal axis); a total area cost of 45 $/$m^2$ (lower axis); a system capital investment of 86 ¢/W_p (left-hand vertical axis); and a levelized busbar cost of 65 mills/kWh (right-hand axis).

To understand the origin of these numbers, note that 1 m^2 of module at 10 % module efficiency generates 100 peak watts (W_p). At 25 $/$m^2$, the module cost is then 25 ¢/W_p. The total area cost of 45 $/$m^2$ is projected as 20 $/$m^2$ for land, wiring, and installation; 17.5 $/$m^2$ for interconnection and encapsulation; and 7.5 $/$m^2$ for metallization and antireflective coating. Nothing has been allocated for the cost of the thin film itself in the minimum cost projection.

The plant efficiency is 90 % and hence 1 m^2 delivers 90 W_p to the grid. The contribution of the area related costs to the system cost is then 45/90 $/$W_p$ = 50 ¢/W_p. To this must be added 14 ¢/W_p for power conditioning. This sum is then multiplied by 1.35 to cover the costs of engineering, interest during construction, and contingencies. Thus the total capital investment is 1.35·64 = 86 ¢/W_p. The fixed charge rate of 15 % per annum requires the system to generate an annual revenue of 13 ¢/W_p to pay the interest, return on equity, taxes, etc. on the capital investment. In the U.S. Southwest the system would on average produce 2.5 kWh/W_p each year, resulting in a levelized electricity cost of 5.2 ¢/kWh or 52 mill/kWh. To this must be added 4 mill/kWh for operations and maintenance and 9 mill/kWh for back-up generation, giving a total cost of 65 mill/kWh.

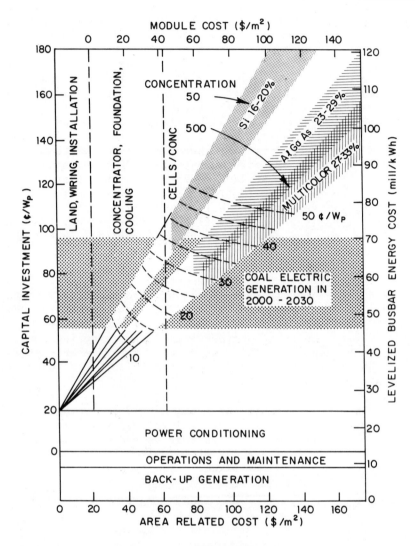

Figure 2

Allowed costs for concentrator PV systems used in
Southwest U.S. central power stations circa 2000-2030

The vertical and horizontal axes and many of the other features are the same as those in Fig. 1. Structural costs are based on the use of enclosed, lightweight, 2-axis tracking concentrators (see Table 5 and text). The optical efficiency of the concentrators is assumed to be 70 %. The module cost includes the cost of the concentrator complete with cells, foundation, and cooling equipment, ready for shipment from the factory.

Three types of concentrator cells are considered: Si, GaAs, and multicolor cells. The efficiency ranges shown for each category are estimated obtainable cell efficiencies at a temperature of 50 C for the indicated concentration ratio of each system.

The demarcation line separating shaded and unshaded portions of each pie-shaped region is based on our estimates of the probable minimum capital investment required for the various concentrator systems and is determined by cell, module, and structure costs (see Table 5 and text). The cell cost contribution to the total system cost per unit aperture area is given by dividing the assumed cell cost per unit area of cell by the concentration ratio; the different cell costs of the three systems determine the minimum area-related costs of the systems, as can be seen in the region labeled CELLS/CONC.

The horizontal shaded band has the same meaning as in Fig. 1. The region marked "back-up generation" shown at the bottom of the figure represents the cost of peaking power which must be added to the PV system to insure a reliability equivalent to the comparison coal station.

The curved dashed lines give the system capital investment for fixed module (or array) cost per peak watt. The lines curve upward for the same reason as in Fig. 1: fixed area costs such as land and installation penalize less efficient systems. Comparison of Figs. 1 and 2 indicates that the noncell module cost per m^2 is considerably higher in a concentrator than in a flat plate system.

    As an example, our minimum projected cost for a 2-axis concen-
trator employing Si cells of 20 % efficiency is the lowest point on
the right-hand edge of the shaded region for Si. Its position
relative to the curved dashed lines indicates an uninstalled array
(or module) cost of 33 $¢/W_p$. Projecting this point onto the four
axes shown on the sides of the graph gives: 47.5 $/m^2 module cost
(upper axis); 67.5 $/m^2 total area cost (lower axis); 91 $¢/W_p$
system capital investment (left axis); and 68 mill/kWh levelized
busbar energy cost (right axis).

    The contributions to the area cost are 20 $/m^2 for land,
wiring, and installation; 42.5 $/m^2 for concentrator, foundation,
and cooling; and a cell cost of 5.0 $/m^2 of collector aperture.
Since the concentration is 50X, the cell cost is 250 $/m^2 of cells.
The optical collection efficiency is 70 % and the cell efficiency
(at the assumed operating temperature of 50 C) is 20 %. Thus peak
insolation of 1 kW/m^2 yields 140 W_p of module output. The cost
of the array per peak watt is then (47.5/140) $/W_p$ = 34 $¢/W_p$.
The plant efficiency is 90 % and thus 1 m^2 of collector delivers
126 W_p to the grid. The system cost per m^2 of module is
67.5/126 $/W_p$ = 54 $¢/W_p$. To this we must add 14 $¢/W_p$ for
power conditioning and multiply the sum by 1.35 to cover the cost of
engineering, contingencies, and interest during construction. The
total system cost is 91 $¢/W_p$.

    With a fixed charge rate of 15 % per annum the annual cost of
capital is .15·91 = 13.6 $¢/W_p$. The system generates an average
yearly energy that is equivalent to operation at peak capacity for
2500 hours. The cost per kWh is thus 5.5 $¢$/kWh or 55 mill/kWh. To
this we must add 4 mill/kWh for operations and maintenance and 9
mill/kWh for back-up generation. Thus the total levelized busbar
energy cost is 68 mill/kWh.

2. It is anticipated that only a small fraction of the elec-
    tricity generated after 1990 will be based on gaseous and
    liquid fuels. Therefore photovoltaics will not signifi-
    cantly reduce the use of these fuels. The major effect of
    photovoltaic generation will be a displacement of some
    combination of coal and nuclear fuels.

PV will be used by utilities primarily to save the fuels used in
conventional electricity generation. Table 6 shows the U.S. generation mix (the fraction of total electricity produced by each source)
existing in 1975 and four predictions of future generation mixes of
the nation's utilities. Over 80 % is expected to be coal and nuclear
generation, about 9 % will be hydro and other renewable sources and
only about 9 % will be liquid and gaseous fuels. The percentage contribution of hydroelectricity and renewable sources decreases because
hydro remains essentially constant in total output power while total
production of power roughly doubles. Very limited production of
solar electricity by the year 2000 is assumed, although other solar
energy sources (e.g., solar heating) may be significant by that time.

A recent study for EPRI[28] which considered fuel displacement
in three representative utilities showed that PV displaced all fuels
used to some extent, but primarily the fuel used in the plants which
meet the increased daytime load, as one would expect. In the future,
such generation will probably be served by intermediate-load coal
plants or by baseload coal or nuclear plants combined with load-
leveling batteries. Therefore, PV will not curb the nation's rising
liquid fuel costs: the value of PV lies in reducing the nation's
future dependence on coal, uranium, and plutonium.

Table 6

Recent and predicted future generation mixes

	U.S.[24] (1975)	Exxon[25] (1990)	El. World[26] (1995)	EPRI[27] (2000)	Aerospace[24] (2000)
Liquid and gaseous fuels	30 %	10 %	9 %	8 %	6 %
Coal	45	54	56	45	54
Nuclear	9	27	28	38	33
Hydro/renewable	16	9	7	9	7

3. Because of the costs associated with encapsulation, foundations, support structure, and installation of PV array fields, there is a large economic penalty for the use of low-efficiency cells. To compete in the U.S. central power generation market, even zero cost PV cells must have a limiting minimum efficiency. The use of modules with efficiency as low as 10% will probably require substantial reduction in these other costs, even if the modules themselves are inexpensive.

Estimates of non-module system costs based on current technology are listed in Table 3 of Section 1.[29] The non-module area related cost obtained is 38.5 $/m^2. In the cost estimates used in Fig. 1 (and in Fig. 2, which applies to concentrator systems), we have reduced this estimated cost to 20 $/m^2, assuming that low cost structures and automated installation methods will be developed. If these developments do not occur, the additional area related cost of 18.5 $/m^2 will increase the capital investment (CI) for a system with 10% efficient flat plate modules by 28 c/W_p and increase the CI of a system with 50 c/W_p 14% efficient Si modules by 20 c/W_p. These systems would then be significantly further from the allowed range of 50-100 c/W_p.

Figure 1 also illustrates the great sensitivity of CI to module efficiency. Under the assumptions used in this study, the minimum value of CI rises above the competitive range as thin film module efficiencies decrease below about 10 %. The requirement of a minimum module efficiency of nearly 10 % is due to the approximately fixed costs per unit module area of land, clearing and grading, wiring, lightning protection, support structures, installation, encapsulation, and metallization. In Tables 3 and 4 these costs are estimated to be at least 20 $/m^2 for the non-module cost and 25 $/m^2 for the minimum module cost, even if the thin film itself is of negligible cost.

4. The minimization of materials usage in the overall PV
    system must be emphasized. New designs for PV systems
    should be tested against the availability of materials at
    high rates of construction of generating capacity, because
    political or economic developments might produce a need for
    accelerated rates of deployment. Some current designs, in
    addition to being extremely expensive, would make major
    demands on U.S. ability to supply certain materials in
    sufficient quantity.

INTRODUCTION

Federal government interest in PV must be based on the assumption that PV will ultimately be capable of being deployed on a scale
sufficiently large that production of a significant fraction of U.S.
electricity or fuel is possible. As an insurance option, the value
of PV is determined not only by the cost of PV systems but also by
the nation's ability to deploy these systems in a reasonable length
of time without excessive demands on raw materials resources and
energy production. Attention must be given to the supply of cell
materials such as gallium, indium, and cadmium, to the materials used
in module fabrication, and to the support materials such as concrete,
steel, and aluminum.

A simple test model, which we will discuss now, examines some
conceptual designs of structures. It should serve as a useful illustration of the questions which must be answered. A similar model
will then be used to examine cell materials. A more detailed study
of materials questions is in progress.[30]

ASSESSMENT OF STRUCTURAL MATERIALS REQUIREMENTS

In order to assess resource production requirements, a
reasonable build-up rate for PV must be assumed. A plausible rate
of US electricity use at the end of this century can be taken as
$4.0 \cdot 10^{12}$ kWh/yr, about twice the 1975 rate. We estimate the
materials requirement to construct new PV plants capable of producing
1 % of that energy and express it as a fraction of current U.S.
production. Three structural systems to support the photovoltaic
devices have been examined. They are:

1. Steel and concrete structures to support flat plates, typical
    of designs now being considered by JPL.[31]

2.   A plastic bubble-enclosed concentrator[32] (1000X) with
     additions to mount and cool the PV cells.

3.   A steel-structure concentrator (20X) proposed to Sandia.[33]

     The PV system efficiencies assumed for converting insolation to
busbar AC power are 10 % for flat plate silicon and 12 % for concentrator systems.

     The results appear in Table 7. Note that the PV resource
requirements are compared with 1974 annual consumption or production,
not with that projected for the year 2000. The structures considered
all use significant amounts of structural material, though no structure is ruled out for the construction rate considered. The advantage of lightweight designs such as the bubble-enclosed concentrator
is obvious, however. If the projections of the utilities industry
are correct, an annual construction rate producing new generation of
about $2 \cdot 10^{11}$ kWh/yr would be required to meet all the marginal
increase in demand for electricity around the year 2000. Meeting
that demand through PV construction would make major demands on some
materials in some designs, as can be seen from the table.

ASSESSMENT OF CELL MATERIALS REQUIREMENTS

     In order to assess cell materials requirements, we again assume
a PV capacity increase of $2 \cdot 10^{10}$ W_p. We assume that Si and
Cu_2S/CdS cells are used for flat plate applications, and that the
cell thicknesses are taken to be 250 microns and 20 microns respectively. The overall system efficiencies are taken to be 10 % for Si
and 7 % for Cu_2S/CdS. We also consider stacked cells for concentrator use (500X concentration) composed of thin active layers of 10
microns of various compounds using In, Sb, and other more abundant
elements. These concentrator cells are assumed to be grown on either
Ge or GaAs substrate single crystals of 300 micron thickness, and to
produce overall system efficiencies of 25 %. The corresponding cell
materials requirements are given in Table 8. The contemplated use of
Ga and Ge is as the substrate material of a stacked cell; of In and
Sb, as one of the constituents of the material in a thin film active
layer in a concentrator cell; and of Cd and Si, as flat plate sheet.

     The requirements shown would permit construction of PV plants to
generate about 1 % of U.S. electricity in the year 2000.

     From this table it can be concluded that a build-up to high
utility penetration by PV would affect the materials supply picture
for each of the cells considered. Ga is available from imported

bauxite ores and could be produced from domestic coal. Costs for
extraction from coal may be prohibitive, however. Ge is present in
coal in about the same amounts as Ga, but extraction cost estimates
are unavailable. Other materials problems for cells may arise from
cell designs which require contact or barrier materials such as gold
and platinum.

Table 7

Structural materials to support and illuminate PV cells.
Materials requirements to add $4 \cdot 10^{10}$ kWh ($2 \cdot 10^{10}$ W_p) of
generation, compared with 1974 U.S. production or consumption.

Array Type	Material	PV Requirement	1974 U.S. Annual Consumption (Production)	Requirement as % of Cons. or Prod.
Flat Plate	steel	$5.2 \cdot 10^6$ T	($1.1 \cdot 10^8$) T	5 %
	cement	$1.4 \cdot 10^7$	($8.3 \cdot 10^7$)	17
Concentrator (steel)	steel	$1.4 \cdot 10^7$		13
	cement	$6.0 \cdot 10^6$		7
	aluminum	$8.6 \cdot 10^5$	$6.0 \cdot 10^6$	14
Concentrator (plastic)	steel	$2.8 \cdot 10^6$		3
	cement	$7.0 \cdot 10^5$		1
	aluminum	$3.3 \cdot 10^5$		6
	oil	$2.5 \cdot 10^6$ Bbl	$6.0 \cdot 10^9$ Bbl	.04

$1 T = 10^3$ kg

The requirements shown would allow approximately 1 % of the
year 2000 electrical generation from PV.

Table 8

Cell materials requirements[a]

Element	$2 \cdot 10^{10}$ W_p Production Requirements	Present Annual Production	Comments
Ge	250 T	76 T (world) 13 (US)	World reserves of Ge are estimated at 2000 tons.[34]
Ga	120	7	Up to 140T/yr could be recovered from Zn and Al refining.[35]
Sb	4	$2 \cdot 10^4$	U.S. use of primary Sb
In	4	76 (world)	
Si	$1.2 \cdot 10^5$	$1 \cdot 10^5$	Metallurgical grade; 1974 production of high-purity Si was about 300T/yr.[34] Estimated 1978 production is 1000 T.[36]
Cd	$2.1 \cdot 10^4$	$6 \cdot 10^3$	U.S. use.

[a]   Because the numbers in this table are based on schematic rather than actual designs, they should be regarded as giving only order of magnitude estimates.

1 T = 1000 kg

5. Energy storage would be necessary if photovoltaics came to
    supply more than about 10-20 % of electrical energy in a
    typical region.

     Storage is not necessary when PV generation represents only a
small fraction of a utility's total capacity. The normal ability of
a utility to change its generation of power to meet changing loads
and to deal with generating failures can compensate for the PV system's intermittent output. The rough coincidence of the PV power
time profile with the daily load curve for most utilities leads to
some capacity credit for the PV system (that is, the addition of a PV
plant reduces the total amount of non-PV generating capacity required
to maintain a fixed total-system loss of load probability).[37]

     In fact, at low PV penetration* system electricity storage and
PV may be in competition, since storage and PV are both sources of
peaking power.[37] On the other hand, storage or back-up is required
when the PV penetration of a utility becomes large enough to under-
mine system reliability because of simultaneous forced outages of all
PV (clouds), or when PV is considered as a source of baseload rather
than peaking electricity for a utility or a source of stand-alone
residential power.

     Simple considerations lead to the conclusion that at 20% energy
penetration storage or fast back-up is necessary with PV. For the
Sunbelt, insolation provides about 2500 kWh/m^2 per year at a peak
level of 1000 kW/m^2. To average 20 % of total generation, PV must
produce about .2·8760/2500 = .7 of the average generation when sunlight is at its peak. This large fraction of system generation is
vulnerable and must clearly be backed up by large spinning reserve or
storage.

*   "Penetration" is often defined as the percentage of a
    utility's maximum generation capacity represented by a
    given energy source. Since PV systems have a low load
    factor (ratio of average to peak power output), this
    definition exaggerates the importance of PV to a utility
    system. We prefer to use "energy penetration", the
    fraction of annual energy output of a utility due to the
    system under consideration. For PV in a normal utility, a
    capacity penetration of 20 % is equivalent to an energy
    penetration of roughly 10 %.

Several studies of storage economics in specific conceptual PV applications have been undertaken.[38] However, no study has, to our knowledge, attempted to deal with the technical problems of the massive amount of storage which PV would require at high penetration.

If PV is to have a truly major impact on the U.S. energy picture it must be technically and economically ready to do much more than make a small penetration in the Southwest. We need realistic goals for PV, storage, and long distance transmission against which to measure progress. These goals should be supplied by national electric system studies which model optimized systems using all power resources and the sharing of PV power between regions.

In order to obtain a rough estimate of the amount of storage needed in a very high PV penetration model, we have used results from a study[39] made for another purpose. This study involved conceptualized homes at seven geographically separated locations in the U.S. Each home was designed to be completely independent of the utility grid and received its electricity from a combination of PV, an oil-fired engine generator, and battery storage. On average, PV and the engine generators supplied about 90 % and 10 % of the primary electrical energy for the seven homes respectively.

A national power system estimate made on this base is at best approximate for a number of reasons. The load curves of the seven houses do not correspond to those of any complete power system, which must also supply industrial and commercial establishments. Our model makes no use of hydroelectricity or the possible advantages of sharing facilities and electricity between adjacent regions. The assumption of 90 % energy penetration by PV uniformly over the nation is extreme in its implication that PV has become much more desirable than all other forms of electric generation. However, the estimate does set an upper bound on storage requirements and develops an intuition for the requirements associated with very high PV penetration.

In constructing the national model, it was necessary to assign a nominal nameplate peak capacity (kW_p) rating to the PV equipment at each location. This was done by assuming an average energy production of 1720 kWh/yr per kW_p of PV. Furthermore, the PV generation energy which was wasted in the isolated house model of Ref. 39, about 16 % of the PV electricity, was taken to be usable in the national model. The characteristics of the individual systems and their sum, which forms the basis for the national model, are shown in Table 9.

Table 9

Annual performance of 7 stand-alone
residential systems,[a] and their sum

LOCATIONS	Atlanta,GA	Cleveland,OH	Madison,WI	Mobile,AL	Phoenix,AZ	S. Maria,CA	Wilmington,DL	Sum[b]
PV kW_p	5.7	6.5	5.2	7.7	5.5	4.4	5.7	40.7
Aux kW_p	1.5	1.5	1.5	1.5	1.5	1.5	1.5	10.5
Storage kWh	20	20	20	25	25	15	20	145
MWh from PV	10.2	8.4	8.5	11.7	15.3	7.4	8.6	70
MWh from Aux	1.1	1.3	0.6	1.3	1.4	0.4	1.1	7.2
MWh total	11.3	9.7	9.1	13.	16.7	7.8	9.7	77.2
kW avg	1.3	1.1	1.0	1.5	1.9	0.9	1.1	8.8

PV kW_p: rated peak output power of PV generation in system

Aux kW_p: "   "   "   "   "   auxiliary generators

Storage kWh: kilowatt-hours of battery storage required in system

MWh from PV: megawatt-hours per year from the PV primary source

MWh from Aux: "   "   "   "   "   auxiliary generators

MWh total: "   "   "   "   "   sum of auxil. and PV

kW avg: average rate of electrical power consumption

[a] Source for systems data: Westinghouse R&D Ctr., Ref. 39

[b] The sum is used as the basis of the national model. See text.

The national system is based on the sum of the seven, given in the last column of Table 9, and has the following properties:

A PV installed capacity of .53 kW_p per MWh of system electrical output each year

An auxiliary generation capacity of .14 kW_p per MWh of system electrical output each year

1.9 kWh of storage per MWh of system electrical output each year, about three hours of storage at system peak capacity

90.7 % of the national system's electricity is associated with the PV capacity and the remaining 9.3 % with the auxiliary generation capacity.

The contributions of storage and auxiliary generation to the cost of each kWh of electricity generated are estimated using EPRI sources. These levelized costs are:

Storage using mid-1980's batteries[40]	.03 $/system kWh
Storage using mid-1990's batteries[41]	.01
Auxiliary generation from liquid-fueled turbines[42]	.01

(Numerical details are given in the footnotes.)

These considerations show the importance of developing economical storage technology if success in the PV program is to be used to affect the U.S. energy picture significantly.

The massive amount of storage needed at high penetration rules out the use of batteries based on lead. The typical lead content of such batteries is 30 kg/kWh. In this model, 90 % PV energy penetration of an electricity system assumed to produce $4 \cdot 10^{12}$ kWh/yr would require $7.6 \cdot 10^9$ kWh of storage, using $2.5 \cdot 10^8$ T of new lead. By comparison, 1973 U.S. lead production was $6 \cdot 10^5$ T and world production was $3.9 \cdot 10^6$ T.[43] U.S. resources are estimated at $1.2 \cdot 10^8$ T and world resources at $3.3 \cdot 10^8$ T. Hence, for large PV penetration another storage system would be required.

INTRODUCTION TO REFERENCES

The references included in this report do not by any means constitute an exhaustive set. They are instead intended to provide entry points to the literature, and for that reason they stress the most recent work.

References which are identified by report numbers without report sources (e.g., SAND77-0909) are reports of the Department of Energy (DOE) or its antecedents such as ERDA. Such reports can usually be obtained from:

> National Technical Information Service (NTIS)
> U.S. Department of Commerce
> Springfield, VA 22161

or

> U.S. Department of Energy
> Technical Information Center
> P.O. Box 62
> Oak Ridge, TN 37830

Other sources of reports cited in these references are:

> Electric Power Research Institute (EPRI) Research Reports
> Center P.O. Box 10090
> Palo Alto, CA 94303

> Jet Propulsion Laboratory (JPL)
> 4800 Oak Grove Drive
> Pasadena, CA 91103

> Solar Energy Research Institute (SERI)
> 1536 Cole Boulevard
> Golden, CO 80401

The frequently used abbreviation PVSC refers to the Conference Records of the IEEE Photovoltaic Specialists Conferences (Institute of Electrical and Electronics Engineers, 345 E. 47 St., New York, NY 10017). Thus the reference "PVSC 13, 1 (1978)" translates as "Thirteenth IEEE Photovoltaic Specialists Conference--1978, page 1". Because the photovoltaic field is so diverse and interdisciplinary, the number of journals containing relevant articles is large. The PVSC proceedings form the most compact source of information on the subject, although the areas of semiconductor/liquid junctions and photoelectrochemistry are not included.

Along with the many books dealing with solar energy in general which contain sections on photovoltaics, there are are several works

on the specific areas of solar cells and photovoltaic conversion.
Some of these are:

C.E. Backus ed., <u>Solar Cells</u> (IEEE Press Selected Reprint
   Series, published by Wiley-Interscience, New York, 1976).
H.J. Hovel, <u>Solar Cells</u>, Vol. 11 of Semiconductors and
   Semimetals, R.K. Wellardson and A.C. Beer eds. (Academic
   Press, New York, 1975).
W. Palz, <u>Solar Electricity</u> (UNESCO, Paris, and Butterworths,
   London, 1978).
D.L. Pulfrey, <u>Photovoltaic Power Generation</u> (Van Nostrand
   Reinhold, New York, 1978).

REFERENCES

1. Our estimate of $5 \cdot 10^{12}$ kWh/year is only an approximate
    number necessary to determine the magnitude of 1% PV
    generation. This estimate is consistent with estimates in
    references 24-27.
2. D. Costello, D. Posner, J. Doane, D. Schiffel, and K. Lawrence,
    Photovoltaic Venture Analysis Final Report Vol. 1,
    SERI/TR-52-040, 1978, Table 11.
3. Aerospace Corporation Energy and Transportation Division,
    Mission Analysis of Photovoltaic Solar Energy Conversion,
    SAN/1101/PA8-1/1-4, 1977.
4. General Electric Co. Space Division, Conceptual Design and
    Systems Analysis of Photovoltaic Systems, ALO-3686-14,
    Vols. 1-3, March, 1977.
5. Westinghouse Electric Corp., Conceptual Design and Systems
    Analysis of Photovoltaic Power Systems, ALO/2744-13, Vols.
    1-5, April 1977.
6. Spectrolab Inc., Photovoltaic Systems Concept Study,
    ALO-2748-12, April 1977.
7. P. Tsou and W. Stolte, PVSC 13, 1196 (1978).
    Bechtel Corp., Terrestrial Central Station Array Lifecycle
    Analysis Support Study, DOE/JPL-954848-78/1, Aug. 1978.
8. General Electric Co. Electric Utility Systems Engineering
    Dept., Requirements Assessment of Photovoltaic Power Plants
    in Electric Utility Systems, EPRI report ER-685-54, June
    1978.
9. J.W. Doane, R.P. O'Toole, R.G. Chamberlain, P.B. Bos, and P.D.
    Maycock, The Cost of Energy from Utility-Owned Solar
    Electric Systems, A Required Revenue Methodology for
    ERDA/EPRI Evaluations, ERDA/JPL-1012-76/3, June 1976.
10. This assumes that support structure costs for central power
    stations can be reduced to the levels used in the cost
    model in the latter part of this Section.

11. Resource Planning Associates, Inc., Institutional Analysis of
    Solar Total Energy Systems, 3rd Quarterly Report,
    ALO/3786-2, Jan. 1978.
    V. Chobotov and B. Siegel, PVSC 13, 1179 (1978)
12. Sandia Project Integration Meeting, Nov. 1978.
    D. Schueler, Sandia, private communication of PV System
    Definition Project results.
13. The capacity factor is the percentage of time a plant is in
    operation. E.A. DeMeo and P.B. Bos in "Perspectives on
    Utility Owned Central Station Photovoltaic Applications,"
    EPRI report ER-589-SR, attribute a capacity factor of 50%
    to intermediate load plants and compare PV systems with
    conventional intermediate load plants.
14. EPRI Technical Assessment Guide, August 1977.
15. Analysis of Alternative Energy Futures, Office of the Assistant
    Director for Systems Analysis, February 4, 1977 (Dept. of
    Energy).
16. W. Vedder, General Electric Co., 1978 (private communication).
17. Aerospace Corp., Aerospace report ATR-78 (7692-01)-1, Vol. 2,
    prepared by Sherman H. Clark Associates (FL spread reflects
    regional variations in coal prices).
18. General Electric Co., Solar Central Receiver Prototype
    Heliostat, Phase I, Executive Summary provided by R.H.
    Horton.
19. D.G. Schueler and G.J. Jones, PVSC 13, 1160 (1978); this paper
    presents a summary of the results of a number of systems
    studies.
20. JPL Project Integration Meeting, August 1978.
21. G.B. Gaines et al., PVSC 13, 615 (1978).
22. D.K. Zimmerman, PVSC 13, 680 (1978).
23. We thank D. Zimmerman of Boeing and R. Alben of GE for helpful
    discussions on this point.
24. Aerospace Corp., Reference 17.
25. Exxon Company, USA's Energy Outlook 1978-1990.
26. Electrical World, "28th Annual Electrical Industry Forecast,"
    Sept. 15, 1977.
27. "Future Sources of Electricity," Speech presented at Edison
    Electric Institute, April 11, 1978, by Floyd L. Culler of
    EPRI.
28. General Electric Co., Reference 8.
29. From Reference 7.
30. J.W. Litchfield et al., A Methodology for Identifying Materials
    Constraints to Implementation of Solar Energy Technologies
    (Battelle Pacific NW Laboratories, Richland, WA 99352),
    July 1978.
    J.N. Hartley, "Photovoltaic Materials Assessment", in
    Proceedings of the Photovoltaics Program Semi-Annual
    Review, Advanced Materials R&D Branch, October 4-6, 1977,
    Golden, CO, CONF-771051, pp. 83-112.
31. Data furnished by W. Stolte of Bechtel Corp.

32. Boeing Engineering & Construction, SAN/1604-1. Vol. 3, p. 4.
33. Spectrolab Inc., Reference 6, p. 7-73 .
34. Bureau of Mines Bulletin 667, Mineral Facts and Problems, 1976.
35. General Electric Co., Reference 8, Vol. 2, p. N-21.
36. Personal communication, Professor Martin Wolf (University of
        Pennsylvania).
37. General Electric Co., Reference 8 .
38. General Electric Space Division, Applied Research on Energy
        Storage and Conversion for Photovoltaic and Wind Energy
        Systems, HCP/T-22221-01/1, January, 1978.
39. P.F. Pittman, Westinghouse Electric Corp. Research and
        Development Center, Conceptual Design and Systems Analysis
        of Photovoltaic Power Systems Final Report,
        ALO/2744-13(Vol.5), May, 1977.
40. From An Assessment of Energy Storage Systems Suitable for Use
        by Electric Utilities, EPRI report EM-264, Vol. 2, July
        1976, pp. 4-48, 4.4.3.1. we obtain:

Lead-acid battery capital cost in 1985	40 $/kWh of storage rating
Renovations to extend life to 20 years	40
Buildings	20
Total capital investment	100 $/kWh of storage rating

At a fixed charge rate of 15 %, the cost of storage per kWh of electricity generated by the total system using mid-1980's batteries is

$$(.0019 \text{ kWh}_s/\text{kWh}_e)(100 \text{ } \$/\text{kWh}_s)(.15) = .03 \text{ } \$/\text{kWh}_e,$$

where kWh_s denotes storage capacity and kWh_e denotes system electrical output.

41. From Reference 37, Vol. 2, p. E41, we obtain:

Improved lead-acid battery in 1995	35 $/kWh$_s$
Lifetime of 30 years with no renovations needed	

Then the cost of storage using mid-1990's batteries is

$$.0019 \cdot 35 \cdot .15 = .01 \text{ } \$/\text{kWh}_e$$

42. From Reference 4, Tables VII-8, VI-3, and VI-5, we obtain the
     following information concerning auxiliary generators:

     Turbine costs

          Advanced combustion turbine   150 $/kW$_p$ rating
          Availability   85.5 %

          Annual O&M (fixed)   .6 $/kW$_p$ capacity
          Annual O&M (use-related)   .00207 $/kWh generated by
                                                           turbine

          Levelization factor used for O&M   1.87

     Fuel costs (assumed heat rate 11 500 Btu/kWh)

          Liquid fuel at 3.6 $/MBtu   .041 $/kWh generated by
                                                           turbine

          .35 % annual real inflation of fuel cost assumed,
          giving a levelization factor of 1.98

With the system data given in Table 9 and on the page following
it in the text, the contribution of auxiliary generation to the
cost of each total system kWh generated is then calculated as

     .00014 kW$_p$/kWh·.15·150/.855 $/kW$_p$ (capital cost)

     + (.6·.00014/.855 + .0021·.093)·1.87   (O&M)

     ı .041·.093·1.98   (fuel)

     = .01 $/kWh

43. Reference 34.

SOLAR ENERGY: THE QUEST AND THE QUESTION

Melvin K. Simmons*

Solar Energy Research Institute
Golden, Colorado

INTRODUCTION

I have borrowed the title for these lectures from a slide show prepared by the Solar Energy Research Institute (SERI) for the International SunDay, 3 May 1978. That was the day of celebration of the great promise of solar energy utilization for human needs. In these lectures, I will review the progress we have made in translating that promise into a reality. The first four lectures will cover solar energy technologies. In the last lecture, I will discuss the basis of solar energy policy formulation and give some of my own views on planning the transition from exhaustable to renewable resources.

These lectures are intended to be introductory. That is, they should provide you an entry into the rich and complex literature now developing in the area of solar energy. I will try throughout these lectures to avoid using the special jargon that has developed very quickly among those who work in this field. However, I am sure at times I will use words or phrases whose meaning will be new to those who have not studied this literature. I will give a number of references which I consider useful sources for further study. However, the field of solar energy is rapidly changing, and so I also provide as Appendix I a list of some journals and newsletters in which you can follow developments as they occur in coming years.

*Present Address: Research and Development Center
                    General Electric Company
                    Schenectady, New York

In the first four lectures, I will be reviewing the various processes and conversion paths by which we can utilize solar energy. There is a great variety in these processes. In order to provide some structure to this review, I first give in this introduction a brief map of the different paths by which solar energy may be converted from its initial resource into the forms in which it is finally utilized (Grosskreutz, 1979). The first major distinction I make is between paths in which the primary process of absorption of sunlight is as heat and paths in which the absorption is by quantum processes. The former I shall call thermoconversion, the latter, photoconversion. In thermoconversion, the energy of the sunlight is instantly degraded in quality to the temperature of a heat reservoir. At this point, the normal laws of thermodynamics apply, including, for example, limits such as Carnot efficiency. In photoconversion, there is no such degradation in the quality of the energy. It remains in a high quality form and can be further converted with high efficiency to electricity, chemical energy, or mechanical energy.

The primary process in thermoconversion paths can occur in either natural or man-made systems (See Figure 1). Within the former, we have the absorption of solar energy at the surface of the earth to produce the evaporation and precipitation cycle of the weather, winds, waves, and the warm surface waters of the tropical oceans. Each of these can be further converted to useful end products including, for example, electricity. Among these many paths, in these lectures I will review only the conversion of energy through winds and the warm surface waters of oceans. I will not review the process of conversion through the evaporation and precipitation cycle which leads to hydroelectric conversion, nor will I review conversion of the energy contained in waves.

When thermoconversion occurs in a man-made system, the solar energy becomes available as heat in a hot solid, liquid, or gas. We can then use this heat directly for heating buildings, water, or industrial processes, or we can convert it into another form by some kind of heat engine. Such engines might use familiar thermodynamic cycles (Rankine, Brayton, etc.) or might use more advanced thermoelectric, thermochemical, or thermomechanical effects. The ultimate product might be direct heat, mechanical power, electricity, or chemical energy.

In solar photoconversion, we can again distinguish between paths in which the initial conversion occurs in natural or in man-made systems (See Figure 2). In natural systems, the conversion is through the biological process of photosynthesis leading to the storage of energy in biological form. In man-made systems, the absorption can occur either in a solid, leading to the familiar conversion path of photovoltaic conversion, or can occur in a liquid

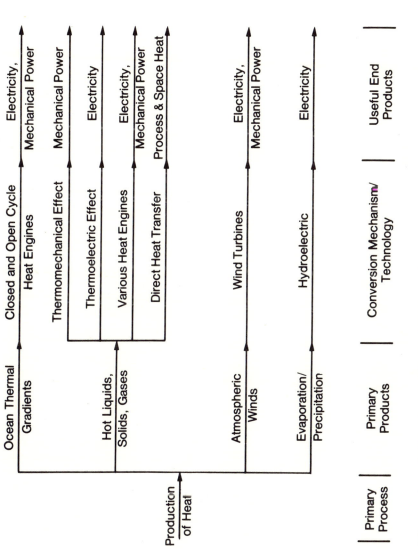

Figure 1.   Paths for utilization of solar energy through thermoconversion as the primary process (Grosskreutz, 1979).

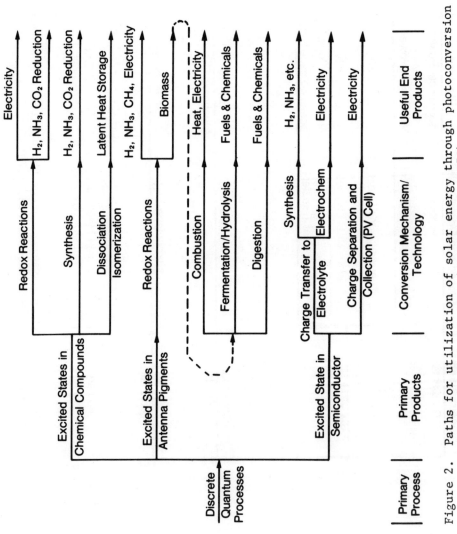

Figure 2. Paths for utilization of solar energy through photoconversion
as the primary process (Grosskreutz, 1979).

or gas leading to the less familiar processes of photochemical conversion.

The pathway of photosynthetic conversion leads to the storage of energy in the form of biomass. The efficiency of this process is generally low, 1% or less. However, it does have the desirable property of yielding an energy form that is easily storable and transportable. Biomass can be burned directly to provide space heat or electricity, or can be further converted by a variety of processes to desired fuels or chemicals. In some cases, biological organisms will naturally convert solar energy to some useful chemical forms, for example, hydrogen, fixed nitrogen compounds, or hydrocarbons.

When the initial photoconversion step occurs in a man-made system, the energy is readily available in the form of electricity or chemical energy. In the photovoltaic cell, the collection is in the form of free charge carriers leading to electricity. If the absorption occurs at the surface of a semiconductor in contact with an electrolyte, then the energy can be available either as electricity or in a chemical species. When the absorption is by a liquid or a gas, the energy can be converted in a wide variety of ways to useful chemical forms for storage or utilization.

These two maps of the pathways for solar energy conversion have introduced quickly the variety of technologies that will be reviewed in these lectures. The complexity of these pathways demonstrates clearly that there are many different ways that solar energy can be utilized. Just as there are many potential technologies, there are many disciplines of scientific and engineering endeavor active in developing solar energy conversion and utilization. Thus, these lectures will provide a quick overview of a large number of fields of active research, each of them worthy of a more detailed review than I can give here.

SOLAR ENERGY RESOURCES, COLLECTION AND CONVERSION

In this first lecture, I will review the nature of the solar energy resources and their measurement, and then give a quick overview of the physics of the collection and conversion of this solar energy with particular reference to thermoconversion.

Insolation

Insolation is the term used to describe the incoming solar energy radiation: the starting point for all solar energy conversion pathways. For our purposes in these lectures, the details of the physical processes occurring within the sun are unimportant. To us, it is sufficient to describe this source of solar energy as being a large black body with the surface temperature of about

6000°K. This body produces a radiation flux at the outside of the
earth's atmosphere with a nearly constant intensity of 1354 watts
per square meter (Watt, 1978). Because of the eccentricity of the
earth's orbit, this insolation outside the atmosphere varies by
±3% during the year. This insolation outside the atmosphere also
varies by ±1% with sunspot cycles. However, both of these variations are minor in comparison to what occurs to the solar radiation
during its passage through the atmosphere. During this passage,
the effects of interest to us occur.

Even on a clear day, much of the sunlight that reaches the top
of earth's atmosphere is either absorbed or scattered before it
reaches the earth's surface. The ozone layer at the top of the
earth's atmosphere absorbs part of the ultraviolet spectrum. An
upper dust layer, air molecules, water vapor, and a lower dust layer
absorb or scatter anywhere from 10%-40% of the incident sunlight on
a clear day. Thus, even though insolation is about 1354 watts per
square meter at the top of the atmosphere, the peak sunlight observed
at the surface of the earth is about 1000 watts per square meter.

Because of these processes, a solar collector pointed up into
the sky will see more than just the light coming directly from the
sun. A collector will typically see four major components of solar
energy. First, the direct beam of light from the sun that is not
scattered in its passage through the atmosphere. Second, it will
see light which has been scattered by small angles and appears to
come from regions near the sun in the sky. This is called circumsolar radiation. It also will see diffuse sky radiation from both
the clear sky and clouds. And, finally it will see light reflected
from the ground or other nearby objects. Types of solar energy
collectors differ in their ability to use these components of solar
energy. For example, highly concentrating collectors can use only
the direct solar energy component. Some types of low temperature
collectors can use all of these components fairly well. Thus, no
single measure of insolation is adequate to describe the amount of
energy available for all types of solar collectors. An adequate
description of the solar energy available at a site can usually be
given by the values of six solar energy components. The total
energy on a horizontal surface is the first and simplest of these.
This is the quantity usually given in tables of insolation data.
More important for most low temperature thermal collectors is the
total energy falling on a tilted surface (tilted south in the
northern hemisphere). Separate measurement is required if this
component is to be known accurately for a certain tilt angle
(Berdahl, 1977). Direct beam energy is necessary for highly concentrating collectors. For collectors with a moderate concentration
ratio, circumsolar radiation may also be important. Diffuse sky
energy is usable by nonconcentrating collectors, and at many sites
can be an important contribution. Finally, radiation scattered from
the ground can be important for flat plate collectors and for passive

solar energy systems. A complete characterization of the insolation resource at a site would include information on all six of these components. However, such detailed information is very seldon available. Much more often, data on only one or two is available, and the others must be estimated by various techniques (Hulstrom, 1978).

Instrumentation is available to measure each of the solar energy components. The instruments now available generally can do an adequate job when new. A major problem is that instruments have not been maintained well over the years in which the presently available data has been collected. Instruments lose accuracy with time and exposure to the elements, and the resulting data may be unusable. This has happened often during the past several decades of collection of insolation data (Berdahl, 1977).

The instrument commonly used for measuring total radiation on a horizontal surface or on a tilted surface is the pyranometer. Pyranometers measure the total amount of radiant energy in a field of view of 180°. Their accuracy can be fairly good, about 3%, though this accuracy may degrade over time. Pyranometers are the instruments most commonly used in insolation data networks such as that operated by the National Weather Service throughout the United States.

The instrument commonly used for measurement of direct solar radiation is a pyrheliometer. This is a much more complex instrument than a pyranometer because it has a small field of view and must track the sun during the day to perform its measurement. The accuracy of such instruments can be quite good, up to 1%. There have been very few of these instruments used in the past, and there is not yet very much data available on direct insolation. Recent installations should provide much improved data within the next few years. This will be particularly important for future installations using concentrating solar collectors.

The discussion to this point has been about the total amount of radiant energy available to a collector. However, also important for many systems is the spectral distribution of that energy. This is important to some degree for all collectors, but is of greatest importance for systems based upon photoconversion: biomass and photovoltaic or photochemical systems. Figure 3 shows the spectral distribution of direct radiation from the sun (Hulstrom, 1978). The top curve is a spectrum as seen outside the earth's atmosphere. It is approximately a black body spectrum corresponding to a temperature of 6000°K. The curves below that one are the spectra of sunlight that has passed through various thicknesses of the earth's atmosphere. Air Mass 1 refers to the spectrum seen when the sunlight comes from straight overhead and thus passes through one thickness of the earth's atmosphere. Air Mass 2 is seen when the sun is at a lower angle in the sky and passes through twice as much

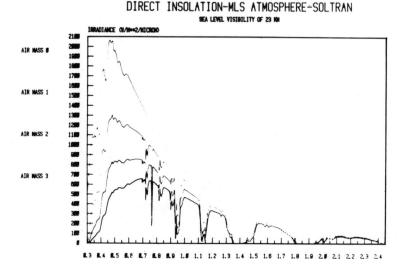

Figure 3.   Spectrum of direct solar radiation on a clear day
            (Hulstrom, 1978).

total mass of air. Air Mass 3 is seen when the sun is even lower
toward the horizon, and three thicknesses of atmosphere are in the
path of the sunlight. It can be seen that both the total amount
of radiation and also the spectral distribution change with air mass
because of the selective absorption of various wavelengths of light
by the earth's atmosphere. This means the efficiency of photovoltaic
or photochemical systems can depend strongly upon the characteristics
of the atmosphere at a site as well as upon the total amount of
radiation available.

     The stochastic nature of the effect of weather on insolation
is important in understanding the performance of solar energy systems. There is no single measure of the randomness of the insolation at a site that suffices to describe the effects of such variations upon all possible solar energy systems. Instead, a simulation
is necessary of the interactions between the solar energy system,
its energy storage, the variations in insolation at a site, and the
load to be supplied. Of particular importance to systems designed
to supply large fractions of a load from solar energy (i.e., high
solar fraction) is the occurrence of long periods of low insolation.
Such occurrences are very site dependent. For example, the number
of times per year that there is an occurrence of a one day duration
of insolation loss is about equal all along the California Coast.
However, occurrences of five days duration are ten times more likely

in the northern part of the state than in the south (Berdahl, 1977).
Thus, assessments of solar systems intended to deliver large fractions of load require careful simulation of the insolation reliability
at a site.

Wind Energy Resources

The resource used in wind energy conversion is the flux of
kinetic energy of a moving air mass. This flux, which has the units
of power, is given by

$$P = \frac{1}{2} \rho v^3 A$$

where

ρ = density of air
v = velocity
A = area intercepted by conversion device.

The factor of v^3 that appears in this equation is crucial to the
nature of the technology used and the problems encountered in wind
energy conversion. Because wind varies greatly in speed from time
to time and from site to site, and because of this v^3 relationship,
the power available from wind energy shows huge variations. These
variations are even more important for wind energy than for solar
energy systems. Indeed, because of this v^3 factor, it is usually
not informative to describe a wind energy system as having a certain
power rating. One must specify also the wind velocity at which that
rating is given. Systems of the same power output can be different
in size by factors of 2 or more if they are designed for sites with
different average wind velocities (Justus, 1978).

The average annual wind power available at a site varies from
less than $100W/M^2$ to more than $500 \ W/M^2$. Thus, wind energy is comparable in intensity to the power density of sunlight but is subject
to much larger variations. Wind power availability varies according
to geography, climate, local terrain, and height above the ground.
I will review each of these items briefly.

Maps have been developed of the large scale geographic variations of wind power in the U.S. and in other nations. These show,
for example, that the mountainous areas of the United States are
areas of high wind energy resource, as might be expected. They also
show that the Great Plains of the U.S. are a significant wind energy
resource. While these maps are interesting, they are of limited
usefulness in estimating the energy actually available at a particular site (Wegley, 1978). This is because the small scale geography
is of vital importance.

Wind energy availability also varies strongly with climate and local weather patterns. Variation with climate is predictable, at least on the average, but the short term variation is not subject to useful prediction.

Small ridges, valleys, hills, or other topographical features cause strong variations in wind energy. For example, when passing over a ridge, the wind speed may double from that at nearby sites on flat ground (Wegley, 1978). From the previous equation, we see that this increases the wind power by a factor of 8.

Wind speed also varies significantly with height above the ground. Wind velocity is zero at the surface and increases in a complex way with height, up to several hundred feet above the ground. The shape of this wind speed profile depends upon the characteristics of the surface and surrounding terrain. This profile is important, not just because it affects the amount of wind energy available, but because it puts great strains on wind devices. In the usual design of wind turbine with a horizontal axis of rotation, blades pass on each rotation from a region of high wind speed to a region of low wind speed and back again. This exerts great dynamic forces on the blade which must be designed against.

Just as wind energy power varies greatly in space from site to site, it also varies greatly in time at a fixed site. The seasonal variation is often quite large. For example, the average wind power available by month can vary by factors of 3 or more (Marsh, 1979a). It is usually, but not always, the case that wind energy is higher in winter than in summer.

The hourly variation of the wind during the day is also important. There is a strong random variation that cannot be predicted accurately. However, in some sites, there is a regular pattern in the daily wind patterns, for example, in certain ocean shore and mountain regimes. Not so important for estimating the energy output of the device, but crucial for its design, are the faster than hourly variations in wind speed. Wind gusts can be both fast and severe and are a challange to the designer of a wind machine. Quite often, data on wind gusts at a site is inadequate, and very conservative design approaches must be taken.

Biomass Resources

The rate of capture and conversion of solar energy by natural photosynthesis is estimated to be about ten times the present use of energy by man (Boardman, 1977). We often speak of this as a biomass resource. However, it is not an inexhaustable resource in the same sense as solar energy and wind energy. Biomass resources are renewable only if they are carefully managed. The need for their management to ensure future productivity will restrict our use of this resource.

Table 1.   World Annual Photosynthetic
Primary Productivity

	Dry Organic Matter (tonnes)	Biomass Energy Content (Joules)	Incident Energy Content (Joules)
Land	10–14 x 10^{10}	1.6–2.2 x 10^{21}	0.6 x 10^{24}
Oceans	7– 8 x 10^{10}	1.1–1.3 x 10^{21}	1.4 x 10^{24}
Total	17–22 x 10^{10}	2.7–3.5 x 10^{21}	2 x 10^{24}

Table 1 gives one estimate of the world annual photosynthetic production of biomass. For comparison, world energy use outside Communist areas will be about 4.5×10^{20} Joules in 1985. It can be seen that although the oceans have a larger area than the land, they produce much less biomass. Indeed, the biomass production on the earth's surface is concentrated in a relatively small fraction of its area. The measure of total photosynthetic activity is usually called primary productivity by ecologists and has been estimated for a number of different regimes and ecosystems (Ehrlich, 1978). Primary productivity is very low, almost zero, in the open ocean, in the desert, and in Arctic regions. It is extremely high in tropical forests and in estuaries and coastal areas. It is not well understood how much of this total primary productivity we might harvest for energy without adversely impacting the operation of the ecosystem. But it is clear that careful policies for management of these resources will be necessary if we wish to use them for production of energy as well as for their traditional roles in production of food and fiber.

Collection and Conversion of Radiation

As a background to our discussion in the next lecture of various solar-thermal conversion systems, I review here the physical processes involved in the collection and conversion of solar energy. Most of this discussion will be directed towards thermal collectors, though some of the principles also apply to photovoltaic or photo-chemical systems. We begin by looking at the processes that occur when incident sunlight falls on a surface (Figure 4). We would like to collect and convert to useful form as much of this incident energy as possible. Thus, we would like to minimize each of the processes of loss of energy shown in Figure 4. First, some of the sunlight will be reflected and not absorbed. To control this loss mechanism, we coat the surface to give it a very high absorbtance. Good surface coatings typically have absorbtance of .9 or higher. However, because the surface has a finite temperature, it will lose energy

by black body radiation. The higher the temperature of the surface,
the higher the rate of loss by this mechanism. We control such
losses from the surface by controlling the emittance of the surface.
Our objective is to achieve a very low emittance for black body
radiation while still maintaining a high absorbtance for the incident sunlight. Energy will also leave the surface by convection.
The ways to avoid this include evacuating the region in front of
the surface or providing barriers that restrict natural thermal convection. Finally, energy is also lost out the back and sides of
the collector by heat conduction. This loss is easy to control in
a good design by proper insulation.

There has been significant progress in the last few years in
developing surfaces that meet our dual requirements of high absorbtance and low emittance. That such a surface is possible is due to
the different spectral distributions of sunlight and of black body
radiation from a collector at normal operating temperature. Almost
all of the energy from the sun passing through an air mass of one
is at wavelengths shorter than 1.5 microns. For collectors at
temperatures up to 700°C or so, almost all black body radiation
emitted is at wavelengths longer than 1.5 microns. Thus, the perfect surface for a solar energy collector would have absorptance of
one for short wavelengths (below 1.5 microns) and zero absorptance
(and thus zero emittance) at wave lengths longer than 1.5 microns.
A number of real surfaces approximate this behavior, though none
do so perfectly. The most common material now used for such a selective surface is called black chrome. This is a specially treated,

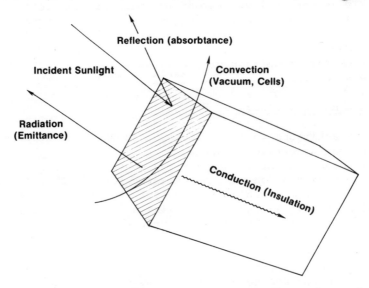

Figure 4.   Processes of energy loss at the surface of a solar energy
            collector and (in parentheses) means for the reduction
            of losses.

do so perfectly. The most common material now used for such a
selective surface is called black chrome. This is a specially
treated, electrodeposited chrome oxide material. It has an absorptance for solar energy of .94 to .96, and an emissivity at a
temperature of 300°C of .05 to .1. Thus, it is highly selective.
This material is now used frequently in commercial solar energy
collectors. A number of such materials are available which have
good properties of absorptance and emittance at low to moderate
operating temperatures. However, there is still a need for development of selective surfaces that operate reliably and for long periods
of time at high temperatures. There is research now underway on
developing such surfaces, and a number of promising new materials
have been identified and are undergoing evaluation (Call, 1979).

Two basically different approaches have been suggested for
control of the convective losses of energy from a collector's surface. The approach considered first to be most promising was
inclusion in the collector of baffles to control the convective
flow of air and thus reduce the rate of heat loss. Typical of such
designs is a honeycomb of a transparent plastic placed in front of
the collector's surface (Kreith, 1978). However, such designs have
shown various undesirable properties, and this approach does not
now seem promising. A more sophisticated way to control convection
is by evacuating the region around the collector. This approach
has been gaining popularity over the past few years. A number of
evacuated tube collectors are marketed now. These collectors show
significantly reduced rates of thermal loss, expecially at high
temperatures. Their costs are still high, but this is sometimes
compensated by their better efficiency in high temperature applications.

The amount of solar energy incident upon the absorbing surface
can also be altered by system design. We can concentrate the incident energy on the surface and greatly increase the ratio of energy
input to the thermal losses by the modes discussed above. Thus,
concentration of sunlight can lead to significantly higher collection efficiencies, especially for collection at high temperatures.

When one thinks of concentrating sunlight, what usually comes
to mind is a lens or a mirror constituting an optical system which
projects an image of the sun upon the collecting surface. However,
not all concentrating systems are of this type. Indeed, the most
efficient concentrating systems are not image-forming at all.
Instead, these systems are designed simply to bring all of the solar
flux incident upon an aperature area onto a much smaller absorber
area. The concentration ratio is then simply described as the ratio
of these two areas. There is a theoretical limit to the ideal concentrating collector (Welford, 1978). For full, three-dimensional
geometries, the concentration ratio cannot exceed $1/\sin^2\theta$ where θ
is the acceptance angle of the concentrator. For two dimensional

concentrators which concentrate sunlight onto a line collector, tha maximum concentration ratio is $1/\sin\theta$. The values of these ratios when θ is the angle subtended by the sun as seen from the earth are 52,000 for three dimensional concentrators and 230 for two dimensional concentrators. The ideal concentrator with this ratio would have a uniform acceptance of sunlight across the disc of the sun and have zero acceptance elsewhere. Such concentrators have been designed. They consist of compound parabolic reflecting surfaces. This geometry is desirable for certain applications, but not for all. The geometry is interesting because it represents the ideal performance which other geometries approximate. In many cases, practical considerations of design, assembly, and cost will dictate that a geometry other than the optimum be used.

A great number of different designs of concentrators have been proposed, constructed, and used. The simple lens concentrator corresponding to the magnifying glass is seldom of practical interest. However, lenses based upon the Fresnel principle are sometimes used for thermal collectors or for photovoltaic arrays. The concentrator design now most commonly used is the parabolic trough collector which focuses the sun's image into a line. These range from fairly low precision collectors of moderate concentrating power to some very high efficiency trough collectors with precise reflecting surfaces that can produce quite high temperatures. Compound parabolic concentrators have also been built and operated for both thermal and photovoltaic conversion systems. For high temperatures, it is generally considered that three dimensional concentration is necessary. Here the traditional design is that of a paraboloidal dish. However, this design has the disadvantage that the entire reflecting surface must move to track the sun. This limits size of each individual concentrator, and in turn requires that systems of large energy output must have many such concentrators.

The concentrating collector design that now seems to offer the best potential for advances in efficiency and economics is the heliostat field with a central receiver. This is a type of Fresnel reflector. The concentrating system consists of a large number of individually directed flat or nearly flat reflective surfaces called heliostats. They are aimed so as to create a stationary image of the sun at a central receiving point. Thus, they use light as the medium of transmission of energy from the field to the central point of energy conversion. This seems to be both more efficient and less costly than designs in which the heat is collected at points throughout the field and piped via a hot fluid or other means to a central conversion plant. The heliostat most often is thought of as a component of a large central station solar thermal plant for production of electricity. However, it is also quite ameanable to application for small scale thermal or electrical processes.

A major challenge for solar energy technology is the development of efficient and economical concentrating systems. In this effort, the heliostat now appears the best contender. Parabolic troughs now cost about $200-$300 per square meter. By 1995, I think that they might be reduced to a cost of $150-$200 per square meter, but I consider it unlikely they will become less expensive than that. Paraboloidal dishes are even more expensive now, about $1000 per square meter, and might come down to about $200 per square meter; but again, I think further cost reductions of this design are unlikely. The heliostat now costs about $250 per square meter, about the same as parabolic troughs. But, I think this design has greater potential for cost reduction. Studies recently completed for SERI suggest that heliostats can be mass produced for about $90 per square meter by about 1995 (Doane, 1979). Because heliostat based systems also have efficiency advantages over trough or bowl systems, this gives them a very strong economic advantage over those designs.

I had planned at this point in the lecture to review the theoretical limits to the efficiency of conversion of solar energy. However, this topic has been covered quite adequately by Henry Ehrenreich in his lectures on photovoltaic conversion, and only a few brief remarks are necessary here. There is a theoretical limit given by thermodynamics for any device that tries to convert a beam of solar energy into another energy form. However, these limits are not usually of practical importance. In thermoconversion, a Carnot limit applies to any heat engine we wish to drive from our collector.   Thus, in practice, the conversion efficiency is limited by the temperature of the collector, and collector materials usually determine a maximum practical operating range. In photoconversion, quite different limits apply. Here, the size of the band gap in a semiconductor (or the exitation energy in molecules) and the reversible nature of quantum processes determines the basic parameters that limit the efficiency of conversion. For a review of these limits to photoconversion, see Hall (1979) or Porter (1976).

THERMAL CONVERSION AND USE OF SOLAR ENERGY

In this lecture, I will review technologies for utilization of solar energy in which the initial conversion step is collection of the insolation as heat on a surface or in an absorbing substance. This will include technologies being developed for thermal conversion to electricity and for the direct utilization of heat in industrial processes and in buildings. I will not review efforts, still in the preliminary stages, toward development of technology for conversion of thermal energy to chemical forms.

Thermal Conversion to Electricity

The development of technologies for the collection of solar
energy as heat and its conversion to electricity has been a major
part of the solar energy effort in the U.S. over the past seven
years. I cannot review here all of the activities that have taken
place and all of the technologies now under development. I will,
however, review briefly a variety of system designs now under consideration for small scale solar electric plants. I also will
describe briefly the project which dominated the U.S. effort in
this area over the past two years, the design and construction of
a ten megawatt solar thermal pilot plant in California. Finally, I
will discuss the newest element of the solar thermal program in the
U.S., the effort to develop technologies for the solar repowering
of existing power plants in the southwestern states.

SERI has recently completed an analysis of the system designs
that could be used for fairly small (1-10 megawatt) solar thermal
power plants (Thornton, 1979). Such sizes might be appropriate for
community, shopping center, or industrial facilities. The different
designs indicated in Table 2 are distinguished by choices among the
options available for the design of concentrator, the process for
conversion of thermal to electrical energy, the form of transport of
energy, and medium of energy storage.

Seven different options for the concentrator were considered
and are listed in Table 2. First is the point focus central receiver design which utilizes the heliostat concentrator described in
the last lecture. In this design, a large number of independently
steered heliostats concentrate light upon a single central receiving
point. Thus, this is a three dimensional concentrating system with
a very high concentration ratio and with the desirable characteristic
that all light energy is received at one place on the site. The
equivalent two dimensional design is the line focus central receiver.
In this design, the heliostats track in only one dimension and direct
the light from the sun to a long line focus. Thus, the receiver must
run across one length of the field of collecting heliostats. Concentration ratios here will be lower than in the point focus central
receiver design. Next is the point focus distributed receiver.
Here the concentration is performed by independently steered paraboloidal dishes which are arrayed across the field. The sunlight is
collected at individual points at the focus of each dish. Thus the
term "distributed receiver" refers to the fact that the solar energy
is received and converted to heat at many individual points throughout the collector field. The fixed mirror distributed focus system
is a two dimensional concentrator in which the mirrors remain stationary and the absorbing receiver must move to follow the moving
focus of the sunlight. The next concentrator is the line focus
distributed receiver tracking collector which uses a parabolic
trough to concentrate sunlight to a line. Here the heat is received
in a long line at the focus of each of many parabolic troughs. Thus,

Table 2. System Options for Solar Thermal
Power Plants of 1-10 MW_e Output

Concentrator Options	System Options												
	Conversion					Transport				Storage			
										Thermal			
	Cent-Rankine	Cent-Brayton	Dist-Rankine	Dist-Brayton	Dist-Stirling	Electrical	Water	Salt	Oil	Electrical	Oil	Salt	Water
Point Focus Central Receiver	●							●				●	
Point Focus Central Receiver		●							●	●			
Line Focus Central Receiver	●							●				●	
Point Focus Distributed Receiver	●							●	●			●	
Point Focus Distributed Receiver			●			●				●			
Point Focus Distributed Receiver				●		●				●			
Fixed Mirror Distributed Focus	●							●				●	
Line Focus Distributed Receiver-Tracking Collector	●								●		●		
Line Focus Distributed Receiver-Tracking Receiver	●								●		●		
Low Concentration Non-Tracking	●								●		●		
Shallow Solar Ponds		●				●	●						●

again it is a distributed receiver. As an option to such a design,
the receiver might move and the mirrors remain stationary. This is
the next option listed. Finally, we also considered two options
without concentration; that is, the concentration ratio is one.
First is a simple flat plate system and second is a shallow solar
pond in which the light absorption is in a large body of water in
which thermal losses are controlled by some means. It can be seen
that these options span a wide range of concentration ratios and
of complexities.

Each of the above concentrator options can be utilized with
a number of different conversion technologies. We selected for
each concentrator option one or two conversion techniques that seem
to provide an appropriate match to the characteristics of the concentrator and its receiver. Thus, for example, with the point focus
central receiver option, we considered both a Rankine cycle and a
Brayton cycle. For the other concentrator options, we considered
versions of the Rankine and Brayton cycles and also considered, in
one case, a Stirling cycle.

The choice of concentrator and conversion options dictates the requirements for the energy transport system used in the facility.   For the various designs considered in the study, we examined transport of energy by electrical transmission and by transport of heat in the form of hot water, steam, molten salt, or hot oil.

Finally, we selected appropriate energy storage options to complete each system design. We considered storage of energy in electrical form and thermal energy storage in oil, salt, and water.

Let us now look at how the system options are combined into complete systems in the two cases of the central receiver concentrator design. The first is a Rankine cycle in which heat is collected at a central receiver in a hot molten salt. The molten salt leaves the receiver at 590°C and is brought either to thermal storage or through a sequence of heat exchangers to produce superheated steam. The steam then drives a turbine-generator set of standard configuration. Energy storage in this system is in the form of molten salt in a reservoir. This form of storage has a number of advantages including the ability to drive the turbine from storage at the same steam temperature and pressure as when the system is operating directly from sunlight. Now consider the Brayton cycle that we might use with the same concentrator option. Here the heat is absorbed at the receiver and heats a compressed gas to high temperature. The gas expands through a turbine, is cooled, and then compressed, and sent back to the receiver. This is a closed cycle Brayton system. This system has some advantages of high efficiency but has the disadvantage that it does not readily lend itself to energy storage at the thermal end of the cycle. Thus, we store energy in this system in the form of electricity. A set of batteries with rectifier and power inverter are provided at the output of the generator. As I will discuss later in our lecture on wind energy, there are serious problems of economics of such dedicated electrical storage. A major advantage of the Rankine design over the Brayton design is its ability to use thermal storage rather than electrical storage.

The objective of the SERI analysis of these designs was to develop a relative ranking of their economics and desirability to users. That ranking resulted in the ordering shown in Table 2. We concluded that for the size range of 1 to 10 megawatts of electric output, the most economical systems would be based upon the heliostat and central receiver design. The system with the Rankine conversion cycle was slightly better than the Brayton cycle. The system designs at the bottom of the list fared vary badly in our evaluation. Thus, at this time, I think it most likely that thermal electric power plants built in the next few decades will be of the central receiver type with various types of conversion cycles.

The largest single project in the United States solar thermal program over the past few years has been the design and construction of a pilot plant at Barstow, California, to prove the concept of central receiver conversion to electricity. This pilot plant was originally intended to prove concepts appropriate for large (100 megawatt or more) solar thermal facilities. However, this pilot plant now is seen also as a proving ground for technologies that might be used in much smaller sizes, such as the 1 to 10 megawatt range previously discussed. The present design calls for a pilot plant that will have an output of 10 megawatts of electricity at peak day output. It will have some storage, enough for about four hours of operation, but only at a reduced power output of 7 megawatts (Brumleve, 1977). This facility has been under design for several years, and construction is starting this fall. The present schedule calls for initial operation of the facility in late December 1981. The total project cost is estimated to be $123M.

There are three major subsystems being developed for the Barstow facility. The collection and concentration system will utilize 1800 individual heliostats, each with about 40 square meters of reflector area. This will be by far the largest array of solar concentrators ever constructed. There is at present a competition between two companies, McDonald-Douglas and Martin-Marietta, for the design and production of these heliostats. One of the indirect benefits of the Barstow program has been the great impetus it has provided towards the development and evaluation of heliostats. These benefits may eventually exceed any we derive from the operation of the Barstow facility itself, especially if heliostats prove to be economical concentrators for other applications.

The receiver subsystem for the Barstow facility has been the subject of great controversy. Based upon what it considered valid commercialization requirements, the Department of Energy selected an external once-through steam boiler for the design of the central receiver. It was considered that this receiver would have much lower mass than the alternative cavity boilers being proposed and that, therefore, the cost of the receiver-tower combination would be significantly reduced (Brumleve, 1979). However, the selection of a once-through boiler is unusual, especially for a system with a heat flux as high as that seen in the Barstow receiver. It has been suggested that there may be problems of dynamic and static instabilities in the boiling process in this receiver. Extensive tests are underway to determine the validity of this concern. It is hoped that this design will prove adequate for the Barstow facility, even if it proves impossible to scale up this receiver design to the larger sizes once envisioned for commercial application of the Barstow type of design.

At the time of initiation of the Barstow project, the major market for solar thermal conversion to electricity was thought to

to be newly constructed stand-alone power plants. However, in the
past few years, another possibility has been widely discussed and
is the basis of a new direction in the solar thermal program. Many
small (50-100 megawatts) oil and gas fired power plants were built
in the southwest United States in the late 1950's. At the time of
their construction, these plants had very low fuel costs because of
the extremely low cost of oil and gas at that time. However, fuel
costs have now escalated rapidly, and U.S. national policy requires
a reduction in the use of oil and gas by electric utilities. Utility
companies have been searching for a way to avoid the loss of their
investments in these plants. One of the possibilities under examination is the conversion of these existing plants to operate from
solar energy for enough hours of the year to reduce significantly
their consumption of oil and gas. There is a high level of interest
among utility companies in the southwest United States in this
possibility.

     The repowering of a power plant would occur this way. First, an
appropriate plant has to be found. The appropriate plant must be in
good working order, with a power output of 50-100 megawatts and with
adequate land at the site or nearby for the construction of a field
of heliostats and a tower and central receiver (Curto, 1978). The
receiver is designed to provide steam of the same quality as that
normally used by the facility when operating from oil or gas. After
repowering whenever sunlight is available, the plant operates with
steam from the central receiver. When adequate sunlight is not
available, it operates from oil or gas. Thus, the value of the
heliostats and central receiver is their fuel savings: they do not
increase the total generating capacity available to the utility.

     SERI was asked to analyze this repowering opportunity and deter-
mine whether or not it represented an important near-term market for
heliostat-based solar thermal systems. Our analysis considered both
the likely cost of heliostat-based systems (the cost of such systems
is dominated by the cost of heliostats themselves) and also the value
to the utility of such a system. This value is based upon the role
of a repowered facility in the overall dispatch strategy of the utility. Thus, we calculated the value to the utility of a square meter
of heliostat surface by estimating the fuel cost it would save. This
value then can be compared to the likely costs of a square meter of
heliostat to determine if this is an economically viable application.

     Figure 5 shows the kind of results found in the draft report
of this study (Doane, 1979). It is seen that during the early 1980's,
the value to a utility of a heliostat exceeds the cost goal for production of heliostats. We now think this cost goal can be achieved.
We conclude that for a period of time, these heliostats would be an
economical investment for utility. However, this conclusion results
from the highly non-optimum composition of the generating mix now
owned by these utilities (they have far too much oil and gas fired

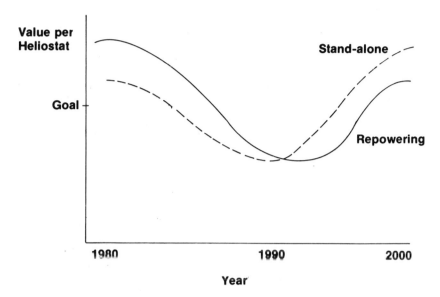

Figure 5.   Value of a unit area of heliostat in a repowering
            application.

capacity). As these utilities add nuclear and coal base-load generating capacity during the next decade, such repowered facilities
more and more often would be providing power in competition with
the fuel costs of coal or uranium. The value of this fuel displacement is not adequate to justify the cost of the heliostat. Thus, we
find that by the 1990's such heliostats will not return value equal
to their cost. In the future, in the year 2000 and afterwards, as
fuels continue to escalate in price, heliostats once again seem to
have a value in excess of their cost. However, at this time, the
greatest value for a heliostat is in a new stand-alone solar thermal
plant, not in a repowered facility. Thus, we conclude that the
repowering opportunity is not a long term significant market but is
rather a temporary opportunity for demonstration of heliostat based
technology. This opportunity should be used as part of the development of heliostat based technologies to be deployed in the year 2000
and afterwards.

Industrial Process Heat

     In the last few years, the United States solar energy program
has increasingly looked towards industrial process heat as a major

market for solar thermal technology. This would involve providing
direct heat for a wide variety of industrial processes, from 50°C
to 800°C from fields of concentrating collectors. It is thought
these applications will generally require full backup capacity by
a conventional energy system. Thus, the value of such systems to
the industry that owns them is only the cost of the fuels they displace. It has been thought that the market for such systems may be
very large because of the huge amounts of energy required in the
United States for industrial processes. However, in market research
we are conducting at SERI, we are finding that the market may not
be as significant as was thought. The underlying problem seems to
be that current industrial processes were not designed to be operated
from solar energy. Rather, they are optimized for other energy
forms, especially natural gas and oil. When one examines an industrial process in detail, it is usually found that there are very
severe requirements upon the form and reliability of energy delivery
for the process. It is very difficult for current solar thermal
technology to meet these requirements. It may be necessary for new
processes to be developed if solar energy is to contribute significantly to industrial energy requirements. These processes would be
designed around utilization of solar and would be well matched to
its characteristics.

The major effort in the solar industrial heat program thus far
has been demonstration of systems in a number of different industries.
For example, moderate temperature solar systems have been demonstrated on laundries and food processing plants. However, these
demonstrations have not given us the indications of economics, performance, and reliability that we would have liked (Mills, 1978).
Costs have generally been very high. The total installed cost of
systems has run $40 per square foot or higher in contrast to the
cost goal of the program of about $5 per square foot. Further,
these demonstrations have not been adequately maintained and monitored to establish their operating efficiency. Data on reliability
are also lacking. Thus, this demonstration program of industrial
process heat has not given us the kind of information we desire from
a new technology demonstration. At this time, I am not optimistic
we will soon have proof of economic and reliable systems for providing solar process heat to industry. What is often considered a
major market may not be of importance to the development of solar
energy over the next few decades.

Energy Use in Buildings

The application of solar energy that most often comes to the
mind of the general public is providing the energy requirements for
residential and commercial buildings. The news constantly reports
on this application, and a number of systems are now being marketed
and installed. It is a major objective of energy policy in the
United States to encourage the adoption of such systems. However, I

do not think the development of technology of these applications
and the design of policies for their encouragement have taken adequate notice of the great changes now underway in the patterns of
energy use in buildings.

The current stock of buildings was designed and constructed
during periods of very low energy cost, and they are extremely
wasteful of energy. The presently high level of energy use for
heating of buildings can be significantly reduced by insulation,
control of infiltration, and passive solar energy design. Other
measures can significantly reduce the amount of energy required
for cooling, hot water, and lighting (Dubin, 1978). These measures
are generally both effective and economical. They have payback
times as short as several months. The adoption of these energy
conservation measurements will have some important consequences
for the effectiveness of solar heating systems for buildings.

The total annual heating requirement for new buildings will
be significantly reduced by design innovations now being developed
and evaluated. A single family residence typical of the present
U.S. building stock requires about 120 MBtu of space heat each year
in a climate typical of the middle United States. In a building
designed with measures that are now economically justifiable, this
heating requirement can be cut to about 50MBtu per year. If further
measures are taken to reduce the infiltration of air into the
building, this heating load can be further reduced to about 30-40MBtu
(Rosenfeld, 1979). At this point, the energy required for space
heating has been reduced to the point that it is comparable with
the energy required for the heating of domestic water. This has
major implications for the sizing of solar energy systems for
buildings: it now becomes just as important to design for the water
heating load as it is to provide for space heat.

Table 3. Free Temperature Rise and
Neutral Point of Residential Buildings

	ΔT_F, °C	T_N, °C
Standard Value	3.9	18
Low Energy Houses		
Arens & Carroll	9.4	11.7
Sinden	10.0	11.1
Kamney	11.1	10.0
Dumont, et al.	20.0	1.1
Robinson	16.7	*

*night set-back

Just as the total annual space heating requirement is reduced
by these design techniques, so are the dynamic characteristics of
the heating load significantly altered. This has important consequences for the interaction of a solar space heating system with
the thermal performance of the house. One representation of this
can be seen in Table 3 (Rosenfeld, 1979). The second column shows
values of the neutral temperature of houses of various designs:
the temperature at which no heating is required. At and above this
temperature the internal heat produced by occupants and appliances
(especially refrigerators and freezers) is sufficient to maintain
the interior temperature at a comfortable level. In houses of
standard current construction, this neutral temperature is about
18°C. In energy conserving houses, this neutral temperature is
reduced to 10°C or lower. This means that the houses no longer
require heating during much of the fall and spring seasons: the
parts of the year during which solar space heating systems are most
effective. For these energy conserving houses, the heating requirement is concentrated in the middle months of the winter season: the
part of the year during which solar energy is least available and
solar systems least effective. Thus, the transition to highly
energy conserving houses can have some adverse effects upon the
match between solar space heating systems and the thermal performance
of buildings. It is important that we design solar space heating
systems appropriate to the types of construction likely in the coming
decades, and not design systems (except for retrofit) for inefficient
buildings that will no longer be built.

Passive Heating Systems

Over the past few years, there has been a major change in our
thinking about how to include solar heating in the design of buildings. We have become much more aware of the opportunities for
integration of the solar energy components with the building structure. In designs commonly referred to as passive systems, this
integration is complete. The building itself serves as the solar
collector and as the solar energy storage. This has economic advantages because the same components serve two roles: part of the
building structure and part of the heating system. This passive
design approach also has the advantage of leading the designer to
a serious examination of the interaction of energy conservation
measures with the performance of the solar heating system (Anderson,
1976).

I will review here five types of passive systems that are now
receiving significant attention. However, this can only provide an
introduction to the great variety of different designs now being
developed, constructed, and evaluated (Mazria, 1979). The five
types of systems I will describe are direct gain, thermal storage
wall, solar greenhouse, roof pond, and convective loop. Most systems
now being constructed fit into one of these categories.

The simplest of these designs to understand is the direct gain system. In a direct gain design, the windows of a house are increased in size on the south face of the building, and the design of the window and the overhang is such as to allow sunlight to enter directly into the living space during the winter months (but not during summer). Solar heat is absorbed on the interior walls and floors of the building and is stored in the mass of the building structure. There is advantage in having large masses exposed to this sunlight. Thus, one sees designs with large masonry walls or floors on the south side of the building. These often lend themselves to very attractive designs emphasizing natural building materials. The openness and airiness of direct gain designs is appealing. However, they are limited in their ability to provide heat to the building without unwanted large variations of room temperature between day and night.

In order to avoid the variations of room temperature seen in direct gain systems, many designers have adopted a thermal storage wall approach. In this design, there are large south-facing glass windows on the south side of the building just a few inches in front of a dark colored heavy wall. Sunlight is absorbed in this wall instead of being passed through to the living space. The wall is composed either of a heavy construction material, such as concrete, or of water in drums or other containers. The heat collected in this wall is then transmitted to the building space through convection and radiation from its interior surface. Because of its large mass, the wall can store heat during the day and release it at night. Such systems have been shown to provide half or more of the heating requirement of buildings in appropriate climate areas. Their disadvantage is that they require a large mass on the south side of the building, and they restrict the amount of free window space on that side. In this regard, they are aesthetically less appealing than direct gain systems, but are more efficient.

The solar greenhouse is a passive design approach of great architectural interest and attractiveness. This design is also compatible with retrofit of some existing buildings. In this design, a greenhouse is built on the south side of the building. The greenhouse is usually built with double or triple paned glass so the heat gained from the sun is not lost to the outside air. Thermal storage is provided by large masses in the greenhouse. This mass can be in the form of masonry or drums of water. Heat is brought into the building from this greenhouse by convection, which can be either natural convection or forced convection driven by fans. When properly designed, such a greenhouse can make significant contributions to the heating requirements of the house. It can also provide a pleasant additional living space for the homeowner. Many such systems have been constructed in New Mexico where sunny, cold winters make this

design especially effective. When the greenhouse is architecturally
well integrated with the design of the structure, this can be a very
appealing design.

    The roof pond system is an approach to passive heating that
also provides significant space cooling in some climates. The
classic example of this approach is the house built by Harold Hay
in southern California. In this design, the house has a flat roof
composed of large water-filled bags or tanks. Panels of movable
insulation ride in tracks above the water bags. This insulation is
moved in accordance with strategies to provide heating in winter
and cooling in summer. During the winter, the insulation is rolled
back during the day, exposing the dark colored bags of water to the
sunlight so the water is heated. During the night, the insulation
is moved to cover the water bags and prevent loss of heat to the
night sky. During these hours, conduction and radiation from the
bags provide heat to the living space. This can be an effective
heating system in mild climate zones. In summer months, the operating strategy of the movable insulation is reversed. During the
night, the insulation is removed from the water bags so that black-
body radiation from the bags to the cold night sky cools the water.
During the day, the insulation covers the bags so that they are
not warmed by the sun. During these hours, the cold water in the
bags keeps the living space cool. In some climate areas, such as
southern California, this system can provide almost all of the
heating and cooling required by a building. Little or no backup
energy is required. However, these systems cannot provide effective cooling in climates with humid summers.

    The last type of passive heating system is the convective loop.
This design uses natural convection to transport heat from an area
of collection to a storage area. This is most commonly used for
heating of domestic water in thermosyphon water heaters. This is
the type of system now commonly used in Israel and commonly installed
in Florida in the 1930's and 1940's. These are efficient and simple
systems, but they require protection from freezing in most climates.
Less common is the use of the convective loop for space heating, but
a few houses of this design have been built. This design approach
is somewhat cumbersome because the process of convective flow requires
that the collectors be placed at a lower elevation than the storage.
This can be done in a house if it is on a steeply sloping site.
Then a collector area with large glass windows is built at the bottom
level of the house. Air heated in this region rises through the
building or through an intermediate area where a large thermal is
provided (for example, a bed of rocks). These are effective systems
but are limited to a few sites.

    This quick tour through the types of passive designs can only
begin to indicate the great variety of approaches possible. We
should expect in coming years to see some of these concepts commonly
applied to the design of most new buildings. The combination of

these passive design approaches and other energy conservation
measures should reduce the heating energy requirements of new
construction to levels far below those typical of our present
building stock.

Active Heating Systems

Active heating systems are the type of solar energy systems
the public hears of most often. There are now between 50,000 and
70,000 active heating systems in use in the United States, more
than in any other nation. Most of these systems are used for heating
swimming pools, but a good fraction, perhaps 20,000, are used for
heating domestic water for residences. A smaller number, perhaps
several thousand, are used for space heating of residences. We are
not sure of the exact number of such systems because there is no
central record kept of their construction. We have only the results
of a few surveys to give us estimates of their numbers. Although
there are many companies now marketing these systems and many consumers have chosen to buy them, their commerical success is not yet
assured. Costs remain high and performance uncertain. While the
numbers quoted above are impressive, they are small compared to the
number of systems that will be necessary if active heating systems
are to displace significant amounts of energy in a modern nation.
That will require millions, rather than thousands, of units. The
commercial prospects of active space heating systems are especially
uncertain. It is yet unclear if such active space heating systems
must compete with, or will be able to compete with, the passive
systems described previously. At this time, it seems the passive
systems may be a more economical and effective approach. After a
passive system is included in the design of a house, the residual
energy requirement for space heating may not be large enough or of
appropriate characteristics to justify the addition of an active
solar heating system.

The basis of most active heating systems is the flat plate
collector (Kreider, 1977). In its most common design, this collector
consists of two covers of tempered glass and an absorbing surface
which often has a black chrome selective absorber coating. Water
is pumped through tubes in this absorber panel to carry away the
collected heat. Such collectors are used in both space heating and
water heating applications, though the water heating applications
are much more common. The heat from these collectors can also be
used, though with great difficulty, to drive a space cooling unit.
Solar driven air conditioners are expensive and relatively inefficient. We have not yet developed an adequate technology for active
space cooling. The belief common a few years ago that space cooling
would significantly enhance the economic viability of an active solar
energy system is no longer generally accepted. Much more effort
is now being directed toward the design of systems that provide only
heat.

Table 4. Reports of Active Solar Energy
Collector and System Efficiencies
(National Solar Data Program)

Collector Efficiency		Storage Thermal Loss (%)	Solar System Efficiency (%)	Solar System COP*	Type of Installation
Predicted (%)	Actual (%)				
48.3	28.0	31.4	19.2	4.7	DHW
43.8	32.0	77.5	7.2	---	Space H & DHW
72.5	43.5	25.0	15.2	47.2	DHW, apartments
47.5	25.0	60.0	10.0	---	DHW,H,C
42.9	25.0	40.0	15.0	1.5	DHW,H,C

* COP total useful solar heat delivered to load/total electrical power for solar operations

There have been a large number of active heating systems installed in the U.S. under the auspices of the government funded demonstration program. Other systems have been installed by home-owners, some with the additional inducement of a tax credit that offsets part of the initial cost of the solar system. Such purchases support several hundred solar energy companies now active in the United States. However, many of these companies are small and are struggling to survive in a period of limited sales.

I would like to report that the systems now being installed are economical and efficient, but that is generally not the case. We have little data on the actual performance in the field of such systems, but the data we do have are discouraging. Table 4 provides data on the efficiencies of a number of systems monitored as part of the United States demonstration program (Ward, 1979). It is seen that the actual solar collector efficiency observed is much less than that predicted. In many cases, it is about half of the predicted level. Particularly disturbing are the large amounts of energy lost from storage. This is shown in the third column. With proper design, this loss should be very small. However, we see here that about half of the energy from the collectors is lost at this point. The effect of these successive losses on the overall system efficiency is shown in the fourth column. It is seen that efficiency is much lower than the 50-60% that would be estimated by the designer of the system. The last column indicates a coefficient of performance for the systems: the ratio of the useful solar heat delivered to the electrical power consumed for operation of the solar energy system (in pumps and fans). This ratio can be very high if the solar energy system is appropriately designed. However, in some cases, the electrical power consumption of the solar energy system is a significant fraction of its energy delivery. With the levels of performance actually seen in the field, active heating systems have very poor economics. The energy from the systems now being installed costs $30-40MBtu. This is not competitive with the cost of electricity or gas in the United States. Thus, I am not as optimistic as many about the potential contribution of such systems in the next few years.

WIND ENERGY, PHOTOVOLTAIC, AND OCEAN THERMAL ENERGY CONVERSION

In this lecture I will review the technologies used for the conversion of wind energy to useful forms and review the techniques employed in assessing the value of wind energy systems. I will then give very abbreviated descriptions of activities in photovoltaic conversion and ocean thermal energy conversion.

Wind Energy Conversion

In this section, I discuss the technologies now available

and being developed for converting the energy resource of wind into useful power. This is a conversion of a flux of kinetic energy of wind into mechanical power and usually then into electrical power. I will begin with a review of the various designs available for this conversion and their efficiencies. These are shown in Figure 6. The efficiency of the conversion of the flux of kinetic energy of the wind to mechanical power output is shown for a variety of designs as a function of tip speed ratio (Wilson, 1974). The tip speed ratio is the ratio of velocity of the tip of the propeller blade to the speed of the wind blowing past the machine. Machines with higher tip speed ratios generally have higher efficiencies. Two designs are shown that operate at a very low tip speed ratio. These are important primarily because they provide high torque, not because they have good efficiency. The Savonious rotor is generally of little interest because of its very low efficiency. The American multi-blade design has been commonly used throughout the American midwest and west for water pumping. This design has advantages of moderate efficiency, very high torque, and excellent reliability. However, it is generally not considered an important candidate for production of electric power (Putnam, 1948).

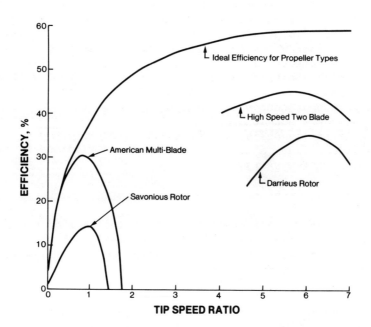

Figure 6.   Efficiency of various designs of wind energy machines (Wilson,1974).

In the conversion of wind energy to electric power, most attention is focused on the high speed, two blade design. This design has a horizontal axis of rotation. It operates with very high tip speed ratios and correspondingly good efficiencies. All very large wind machines including the recent demonstrations constructed by the Department of Energy are of this type. The Darrieus rotor is less common but has some promising characteristics. This design uses two or three blades rotating about a vertical axis and thus accepts wind from any direction. This helps make it immune to effects of wind gusts. The vertical axis of rotation also allows the generator to be mounted close to ground level, which can reduce the size and cost of the tower. Tests of this type of machine are now underway. Some proponents believe it will prove more advantageous than the horizontal axis two blade type.

There are also a number of advanced designs under development with funding from the Department of Energy. There are a great number of geometries that can be imagined for conversion of the kinetic energy of wind into useful power. Some of the geometries may have important advantages. However, the ultimate test is whether an economical machine can be built based upon the design. It is not yet clear that any of the advanced concepts will have a significant advantage over the horizontal axis two blade design or the Darrieus rotor design. In any case, the development of an advanced concept is not necessary for the commercial utilization of wind energy. The existing designs are adequate for our purposes.

It will be noted in Figure 6 that the ideal efficiency for propeller type machines approaches 60% at high tip speed ratios. It is simple to demonstrate, using only conservation of momentum and the Bernoulli equation, that the maximum efficiency possible for this type of machine is 59.3%. It is possible to design machines that approach this level of efficiency. Efficiency is not the major challenge facing the designer of a wind energy machine. The major challenge is the environment within which the machine must operate.

The operating environment of a wind machine is harsh and presents a number of hazards to the machine. One of these, mentioned earlier in the lecture on wind energy resource, is the differential in speed that exists between lower and higher elevations above ground. As the blades of a horizontal axis machine rotate, they pass from a regime of low wind speed close to the ground to a regime of much higher wind speed at the top of their rotation. This exerts strong dynamic forces on the blades, the rotor, and the tower. The machine must also withstand turbulence in the wind stream which deviates significantly from the smooth flow patterns we would desire. At times, the machine will be exposed to winds of extremely high velocity which it must somehow survive. It will also be threatened during its life by lightning, large hail stones, and ice. It must be designed to withstand all of these for a useful operating period

of twenty years or more. These requirements constitute severe
challenges to the designer of a wind machine, and it is this, much
more than efficiency, that concerns those now working in the development of wind energy machines. Indeed, I have heard wind energy
machines described as being a material fatigue testing device which
happens to provide power as a sideline.

Since I have referred a number of times to the conversion
efficiency of wind machines, I should hasten to add that they are
not devices of constant operating efficiency. Rather, a much more
complicated specification of their performance is necessary
(Golding, 1976). At a minimum, the following information must be
given to define the operating properties of a wind machine. First,
there is a cut-in velocity below which a machine has no power output. As the wind reaches this cut-in velocity, the machine will
start to produce power. Power output will increase steadily as
wind speed increases until it reaches the power rating of the generator. At this point, the blades of the wind machine are adjusted
to maintain the power output constant (at the rating of the generator) if wind speed increases further. Thus, at high wind speeds,
the efficiency of the machine decreases so as to keep power output
constant. Finally, there is a wind cut-out velocity above which the
machine cannot operate. At the cut-out velocity, the machine goes
into a protective configuration (blades folded back in some way)
which is intended to allow it to survive in extremely high winds.
In this condition, its power output is zero. Thus, the specification of at least these parameters is necessary if one is to estimate
the performance of a certain wind machine at a site: the cut-in
velocity, the power rating, the velocity of wind at which the power
rating is achieved, and the cut-out velocity (Justus, 1978).

Value of Wind Energy Conversion

Estimation of the value of the power from the wind energy
system is a complex process that raises a number of interesting
issues. I will review those issues in the context of utility connected machines. That is, I will not discuss the value of machines
for remote applications where there is no connection to a utility
grid. Such applications will be of little importance in contri-
buting to the energy budget of a nation such as the United States
where the grid development is already extensive. My discussion
will hold both for cases in which the utility owns the machine and
in which the consumer owns it and sells power back to the utility.
There is no important difference between these cases for the points
I will make here. It is required only that the power output of the
wind machine be supplied to a utility connected application.

The value of a wind energy system derives from the total impact
of the wind machine on the cost of operation of the utility. These
costs include fixed charges, fuel cost, start-up cost, operation

and maintenance costs, and fuel inventory costs. Each of these
items will be impacted by the addition of a wind machine to the
utility grid. The impact upon the first two is most important.
The impact upon the fixed charges of the utility is referred to as
the capacity credit of the machine. The impact on fuel cost comes
from the reduction of consumption of various fuels that is made
possible by the availability of the machine to the utility in meeting
its total demand for energy.

The capacity credit of wind machines has been a complex issue
which is not well understood. First, let us define what constitutes
capacity credit. A utility adds generating capacity to its system
in order to maintain the system reliability, referred to as the loss
of load probability (LOLP) at a required level (Kahn, 1979). As the
demand on the utility from its consumers grows over time, the loss
of load probability will increase unless the utility adds generating
capacity to the system. The addition of a unit of generating capacity will reduce the loss of load probability faced by the utility,
allowing it to serve a certain increase in power demand by its consumers. The capacity credit of the new unit is the increase in
demand which can be met at a fixed loss of load probability. Because
it is a probabilistic concept, there is a non-zero capacity credit
for any unit added to the utility, no matter how unreliable it is.
Thus, a wind machine will always have a capacity credit greater than
zero.

Estimation of the capacity credit requires a simulation of the
operation of the new unit on the utility grid. An hour-by-hour
simulation of the operation of the utility over an entire year is
usually necessary. Such simulations have been done for hypothetical
wind energy machines operating on a number of utility grids in the
United States (Marsh, 1979). From these simulations, several conclusions can be made. The capacity credit of wind machines is sometimes quite good, up to 40% of the rated power output of the wind
unit. (The capacity credit of a conventional power plant is 60-70%
of its rated output.) However, the capacity credit of a wind
machine is often much lower than 40%. In some utilities, where
there is a particularly poor match between the times of high wind
speed and the times of peak demand on the utility, the capacity
credit is near zero: values of a few percent may be found. These
simulations also show that the first wind machine added to a utility
grid has the highest capacity credit. Each subsequent machine will
have a lower capacity credit. For most utilities, the capacity
credit will become quite small, only a few percent, when wind
machines constitute about 10% of the total generating capacity of
the utility. Thus, utilities probably will not install wind machines
constituting more than about 10% of their total installed generating
capacity. This is an approximate upper limit upon the penetration
into utility connected markets of wind machines.

The major value of a wind machine to a utility is the fuel
savings it makes possible by its operation. In circumstances of
correlation of high wind speeds with peak demand on the utility,
these fuel savings can be significant. In most utilities, peak
loads are met by inefficient units which burn oil, the highest cost
fuel used by the utility. Thus, wind machines have high fuel credits
in this type of utility application. In this regard, we should note
the importance of utility energy storage to the value of wind
machines. If the utility has adequate energy storage available to
meet peak loads, then the wind machine has value only for displacing
low cost fuels, such as coal or uranium. This is why wind machines
will have poor value to utilities that have a high proportion of
hydroelectric capacity on their grids.

This adverse effect of the availability of electrical storage
on the economics of wind energy also applies to solar energy devices
connected to utility grids. This represents one way in which the
development of advanced storage technologies may hamper rather than
aid the utilization of solar energy. This problem will occur whenever the storage is nondedicated, that is, can be recharged from
any generating unit on the grid. In this case, the utility will
choose to charge it from its lowest energy cost unit, probably a
large coal burning plant or a nuclear reactor. If the storage is
dedicated, that is, it can be charged only by the wind or solar
facility, then it will increase the value of the wind or solar
facility to the utility. However, nondedicated storage will almost
always decrease the value of wind or solar units to the utility.

By examination of simulations of the operation of wind machines
on utility grids, we can derive an estimate of the total value of
the wind machine to the utility from its capacity credit, the present
value of its fuel savings, and other benefits to the utility operation. There is great variance in this value. In advantageous
circumstances, the value of the wind machine is as high as $700 per
rated kilowatt of output. In other circumstances, the value is
near zero. It is also true that the first units installed by the
utility have the highest value, and the value of each subsequent
unit decreases rapidly as the amount of wind energy connected to
the grid increases. It is hoped that the units now under development in the U.S. program can be produced in large quantities at a
cost of $700 per rated kilowatt. If this cost goal can be attained,
then I expect that we will see these machines purchased and installed
by utilities, at least up to the level of about 5% of their generating capacity. This would constitute a significant contribution
to U.S. energy requirements. However, in this market, wind energy
machines face stiff competition from load management, nondedicated
electrical storage, and new technologies for small conversion units
of low fuel cost, such as advanced coal burning plants.

Photovoltaic Conversion

I will provide only some brief comments on photovoltaic conversion because this topic is being covered so well in the lectures in this course given by Henry Ehrenreich. Indeed, I only include mention of photovoltaics because I would be uncomfortable giving a series of lectures on solar energy without including it, however briefly.

The current state-of-the-art in photovoltaic conversion is represented by the single-crystal silicon cell. Arrays built up from these cells are now sold commercially throughout the world for remote power applications. They cost about $20-$40 per watt of peak output, and about a megawatt of peak output is sold each year (Costello, 1978). This is a strong and viable commercial market which is growing rapidly. However, this current state-of-the-art has little relevance to the kind of technology needed if we are to have large energy contributions from photovoltaic conversion. The single-crystal silicon cell is contructed by a complicated, laborious, and expensive process. Much of the starting material is wasted in the process, and large amounts of labor and energy are consumed. This is what puts the price of these cells in the $20-$40 price range. As discussed by Ehrenreich is his lectures, if we are to see commercial applications of photovoltaics in the U.S., we will have to reduce the cost of arrays to about 20¢-40¢ per peak watt (Ehrenreich, 1979).

At this time, we have only vague ideas about the kind of technology that might produce photovoltaic cells at prices this low. It must be some sort of continuous, fast production system that makes efficient use of its starting material. It should have low labor and energy requirements in order to keep production costs at a minimum. With all of this, it must still produce a cell of high efficiency, since cells of very low efficiency cannot be economically used no matter how cheap they are. The most important activities in the photovoltaic program at this time are research projects working towards development of advanced cell types. We can be hopeful that within a few years, these concepts will come to fruition with designs for low cost, high efficiency cells.

In advance of the availability of that low cost technology, the Department of Energy is funding a program of application tests of photovoltaic devices. Through this program, we are gaining experience with the real world requirements for photovoltaic systems. This is an important step toward the realization of a commercial photovoltaic technology. However, we should not think that the construction of such application tests is any indication that we now have a commercially viable photovoltaic technology. The cost of these demonstrations is exceedingly high and is not justified by their power output. Rather, their value is the information provided for guiding the photovoltaic research and development program.

During the years until we have available a low cost photovoltaic technology, the commercial photovoltaic industry will probably look to the developing nations as its major market. In nations where there is no electric power available in rural areas, photovoltaic systems can provide great benefits to the inhabitants. I expect to see a large program of installation of such systems in coming years, funded by bilateral and multilateral aid agreements. This market will probably keep photovoltaic industries viable during the years before the development of low cost technology will finally allow them to sell devices in utility-connected applications in developed nations.

Ocean Thermal Energy Conversion

This topic I will also cover briefly. This is not because it is discussed elsewhere in the course, but because I am personally somewhat pessimistic about the potential of this technology. I will review briefly the nature of the energy resource and the two generic types of technology under development: closed cycle and open cycle.

The ocean thermal energy resource is the warm water of the tropical oceans. Over about 20% of the ocean's surface, the waters average 25°C or warmer. This resource is located in a broad band extending on both sides of the equator. This is an energy resource because the deep waters at the same sites are much colder, typically 4-5°C. From this temperature difference, energy theoretically can be derived. It is the huge amount of heat in this resource that causes the great interest in its development. It also has the advantage that this resource is available day or night, summer or winter, and does not suffer the fluctuations typical of other solar energy or wind energy resources. Thus, it can provide reliable base load power. Unfortunately, the location of the resource is distant from most sites of significant electrical power demand. Thus, long distance transmission of the energy in electrical or chemical form may be necessary for significant use. A few islands, notably Puerto Rico and Hawaii, are located in areas of significant ocean thermal resource. Since these islands pay very high prices for the electricity that they generate from imported oil, they are at least potentially markets for ocean thermal energy conversion.

The type of conversion technology that is the target of most of the U.S. program is the simple closed cycle Rankine engine (DOE, 1978). This closed cycle engine would be operated between temperatures of the surface waters and the deep waters. However, because of the small temperature difference, only about 20°C, the Carnot efficiency is quite low. The overall efficiency of such an

engine would probably be about 2%. Operation of the engine requires
water to be pumped from far under the surface, 600 feet or so, and
also from the ocean surface around the plant. Significant energy
will be required for operating these pumps. Thus, the net power
output of the facility will be less than its gross power output.
Indeed, until recently, we did not have demonstration that it is
actually possible to operate an ocean thermal facility at a positive net power output. Within the last year, a small ocean thermal
machine (called mini-OTEC) was operated off the coast of Hawaii and
succeeded in demonstrating a positive net power output. This is
particularly impressive because the net output was achieved on such
a small machine. It is expected that the percentage of gross power
appearing as net power will be much better in larger devices.

A number of serious design problems confront ocean thermal
energy machines. These are all made difficult by the very low efficiency of the conversion engine. First, the devices must be very
large and move massive amounts of water. Second, there must be
highly efficient heat exchange between hot and cold waters and the
working fluid of the engine. This means that heat exchanger surfaces must be kept quite clean for long periods of time in an ocean
environment. This is not trivial. Third, the machine must survive
for long periods of time in a harsh ocean environment where it is
subject to storms, strong currents, and wave action. Finally, the
energy produced must be transported to a site where it finally can
be utilized. We do not yet know the most effective way to achieve
this energy transport, though some solutions may be at hand.

As an alternative to the closed cycle design, there is some
research underway on advanced concepts for ocean thermal conversion (Shelpunk, 1979). One of these is the open cycle engine. In
this device, the working fluid is the sea water itself. Water
vapor that evaporates (under partial vacuum) from warm ocean water
passes through a large turbine to condense on the surface of colder
water brought up from the depth. This design avoids the problem
in the closed cycle of heat transfer between the ocean waters and
the working fluid. However, it has its own problems. It requires
a very large vacuum chamber and a turbine that operates efficiently
with vapor at less than atmospheric pressure. The turbine in an
open cycle ocean thermal plant must be sized even larger than the
blades of a wind energy turbine. At this time, we must consider the
open cycle concepts for ocean thermal conversion as being quite
speculative.

Tests to be conducted by the Department of Energy during the
next year on a floating test bed will establish some of the important
design parameters for the closed cycle ocean thermal design. This
will start to give us reliable engineering information that will
allow careful assessment of the feasibility of this technology.

AGRICULTURE, BIOMASS CONVERSION, AND PHOTOCHEMISTRY

The primary process in production of biomass is photosynthesis.
In this process, sunlight, carbon dioxide, and water are converted
into energy-rich chemicals. Whether the reaction takes place in a
bacterium or in a tree, the result is called biomass. For out purposes, we will distinguish between two types of biomass production.
In the first, growth of the organism is assumed to have already been
occurring, either as natural growth in the wild, or as part of agriculture. We then intend to divert some of this biomass (usually
wastes or residues) to an energy product. In the second type of
biomass production, we set out to cause plant growth that would not
have otherwise occurred. This approach we will refer to as energy
farming. In this case, new resources, especially land, water, and
nutrients, are required.

Whether the biomass we have obtained is a residue from ongoing
agriculture, or is obtained from energy farming, we have a number
of options for converting the raw biomass into an energy product.
Two basic categories of conversion processes exist: wet bioconversion and dry thermoconversion. The wet bioconversion processes
include chemical extraction, anaerobic digestion, fermentation, and
distillation. In extraction, chemical techniques are used to extract
valuable hydrocarbons or other chemical forms from the plant material.
An example of this is the isolation of rubber or various resins from
plant stalks. In anaerobic digestion, the usual product is methane.
In fermentation and distillation, the intermediate product is usually
sugar with the final product usually ethanol.

Thermoconversion processes include gasification, pyrolysis,
liquefaction, and combustion. Gasification can occur with air, the
result being a low energy content gas, or with oxygen, the result
being a medium energy content gas which can be used directly or converted to methanol or ammonia. In pyrolysis, the usual products are
oils, gases, and charcoal. In liquefaction, the products are various
mixtures of oils and gases. In combustion. the product is direct
heat energy which might then be converted to electricity.

Table 5 tabulates by type biomass facilities in operation in
the United States. This sample is taken from a recent review of
biomass conversion (Klass, 1978). It should be noted that a great
variety of conversion techniques (listed along the left edge) are
being investigated, as are a wide variety of feedstocks (listed
along the top). It should also be noted that even for each feedstock, it is not clear yet which conversion technique is best.
Indeed, research on a number of conversion techniques is taking
place for each possible feedstock.

The reason for the diversity of conversion approaches for each
feedstock is our present uncertainty as to the most valuable form

Table 5. Examples of Biomass Pilot, Demonstration, and Commercial Plants in U.S.

	Municipal Solid Waste	Industrial Waste	Refuse Derived Fuel	Forest Residues, Wood, Waste Cellulose	Manures	Other
Separation	25	–	–	1	–	–
Combustion	22	1	5	8	–	3
Landfill	5	–	–	–	–	–
Anaerobic Digestion	5	1	1	–	14	3
Fermentation	1	–	–	1	–	2
Liquefaction	–	–	1	–	–	–
Pyrolysis	6	2	2	1	–	2
Partial Oxidation	6	1	5	–	–	1
Other	4	–	–	2	2	–

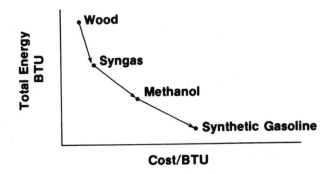

Figure 7. Sequence of steps in converting biomass to higher
           value fuels (Smith, 1979).

for biomass energy. Figure 7 shows the sequence of steps that may
be taken in upgrading biomass to other energy forms (Smith, 1979).
In this example, wood is first converted to a synthetic gas, which
is then converted to methanol, which is then converted to synthetic
gasoline. Each step in the process imposes a cost in equipment and
operations. Further, each step in the process is less than 100%
efficient. Thus, as we move along this sequence, the total energy
contained in the fuel decreases and the cost per unit of energy
increases. It is an open question as to which step along this conversion path has the best ratio of cost to value. It is generally
true that the more processed forms (e.g., synthetic gasoline) have
higher value in today's markets. However, it is costly to convert
biomass into those forms. Biomass may have a greater value in
future markets in its less processed forms, such as synthetic medium
energy content gas.

Agriculture and Biomass Issues

The production of biomass for energy will impact agriculture in
a number of ways. Most obviously, biomass production competes with
agriculture for resources, especially land, water, and nutrients.
However, in some circumstances, biomass production may also be com-
plementary to agriculture and allow greater net benefit to be derived
from agricultural production activities. We do not yet know to what

degree agriculture and biomass production will compete or will complement each other. However, it is clear that the development of
biomass will be intimately linked to national agricultural policy.
A number of important questions about this linkage come to mind
(Flaim, 1979). To what extent may subsidies to agriculture (about
$5B in the United States in 1978) be redirected from limiting food
production to stimulating production of energy crops? What will be
the impacts of producing energy crops on the prices paid by consumers for food and fiber? What will be the consequences of a major
crop failure if food and crop residues have been committed to energy
production? Can crop residues be harvested for energy without
depleting soils and ground water? These and other issues will
certainly be debated in coming years. Research to clarify the nature
of these issues is vital.

I will discuss here only one of these issues, the last question
posed above about the use of crop residues. Much of the energy produced in the growth of a plant remains in the residues left in the
field after the food part of the crop has been removed. The naive
conclusion is that the residue is waste which is clearly available
for any beneficial use such as energy production. However, such
crop residues when left in the field play an important role in
maintaining the quality of the soil and its ability to produce crops
in future years. Thus, there is some constraint on our ability to
collect crop residues for energy. Nutrient depletion and soil tilth
are impacted because removing the crop residues can deprive the soil
of needed potassium, nitrogen, and organic matter called tilth,
which is required for good soil productivity. Water infiltration,
evaporation, and soil productivity are all improved by the crop residue remaining on the ground. Finally, soil loss by water and wind
eroison is reduced by the presence of crop residues on the ground.

After consideration of these factors, some analysts have concluded that very little, if any, crop residues can be converted to
energy without adverse effects upon agricultural productivity.
However, recent research suggests the conclusion need not be this
negative. Experiments which have used good techniques for handling
crop residues and for maintaining soil condition have shown that it
is generally possible to take about half of the crop residues from
the land without any significant adverse effects on future soil
productivity (Flaim, 1979). These results, if true generally for
the nation, indicate that a few percent of the nation's total energy
requirements could be provided from proper use of crop residues.
While this research was done on farms in the United States and is
directly applicable only to U.S. agricultural practice, it is probably true that most developed nations can derive several percent of
their energy requirements from crop residues.

Alcohol Fuels from Grain

A highly complex policy issue regarding the interaction of
agriculture and biomass energy production is now being debated in
the United States. This debate concerns the conversion of excess
grain to alcohol for the operation of automobiles fueled by gasohol.
Gasohol is a mixture of 90% gasoline and 10% ethanol. There is a
special opportunity for production of gasohol in the U.S. because
of our surplus production of grain. The conversion of this grain
surplus to gasohol is being supported by farm interests in the U.S.
They believe such production will help stabilize grain prices from
the wild oscillations of the market.

Among the complex issues raised in the gasohol controversy are
a number concerning grain exports from the U.S. Such exports provide
the United States with an important flow of foreign currency. Funds
gained from grain exports allow the United States to purchase large
amounts of foreign commodities, including OPEC oil and other raw
materials important to the U.S. economy. However, this grain is even
more important to the nations that require it as a supplement to
their own food production. The conversion of this food stuff into
alcohol to drive American cars could become a major international
issue if world-wide grain shortages develop in the future.

The economics of the fermentation process usually used for
production of ethanol from grain are not very advantageous. About
2.5 gallons of ethanol can be produced from a bushel of corn. In
metric units, this is about 37 liters of ethanol from 100 kilograms
of corn. A byproduct known as distiller dryed grain (DDG) is
obtained and contains much of the biomass from the initial feedstock.
Because of the low value of this byproduct, ethanol produced by this
route is not competitive with the cost of gasoline in the United
States at this time. However, it is possible that alternative processing techniques can be developed in which more valuable byproducts
could be derived. For example, one corn processing technique would
result in production of significant amounts of corn oil as well as
the same amounts of ethanol as the normal fermentation approach
(Hertzmark, 1979). The high value of the corn oil could significantl
offset the cost of the ethanol.

Part of the debate about the gasohol production in the United
States has been about the net energy or net liquid fuel balance of
the program. As I noted above, about 2.5 gallons of ethanol are
obtained from each bushel of corn. However, a number of analyses
indicate that it generally requires more energy than this in gasoline
and other petroleum products to produce that bushel of corn and convert it to ethanol. Thus, by these analyses, each gallon of gasohol
produced actually imposes a net loss of liquid fuels and a net loss
of energy to the United States. Some have suggested that this negative net fuel balance can be corrected if non-petroleum fuels (coal

or biomass) are used for the production and conversion of the corn.
We might then have a net energy loss from the gasohol program, but
a net gain in liquid fuels. Such an arrangement seems possible, but
we have yet to formulate the public policy tools that will assure a
net gain of liquid fuels from a gasohol program.

Given the many uncertainties, it is difficult to estimate the
actual impact of a gasohol program on oil imports into the United
States. One must consider the loss of foreign revenue by a decrease
in grain exports, the potential value of exports of byproducts (such
as a DDG) from gasohol production, and the net liquid fuels balance
of the gasohol production. The debate on this issue is far from
settled. However, strong political interests assure that the decision whether or not to proceed with a large gasohol program will be
made on many grounds other than just the efficient production and
use of biomass energy.

Digestion of Solid Wastes

Unlike the complex case of agricultural residues discussed
above, there seems to be nothing but benefit to be derived from
increased utilization of municipal solid wastes (garbage to most
of us). The severe problems of properly disposing of these wastes
provide a great inducement to finding better ways to process these
materials and derive any possible benefit from the resources they
contain. Much of the value of these resources is in metals and
other materials. However, municipal solid wastes also represent
a significant energy resource. A number of projects are now underway to demonstrate the conversion of these wastes to useful energy
forms (Anderson, 1977). One of the approaches being studied is
anaerobic digestion of solid wastes to produce methane, a process
already widely used in Europe.

In this process, the municipal solid wastes from a community
must be collected and brought to a large conversion facility. There
the raw waste is finely shredded and is air-classified to separate
light and heavy components. The heavy components contain valuable
metals and other materials. The light fraction is largely various
forms of biomass and plastics, all of it energy rich materials
(Goddard, 1975). In the anaerobic digestion approach, this material,
after being blended and scrubbed, is fed into a large digestion tank.
Here microorganisms slowly digest the waste and produce a gas consisting primarily of carbon dioxide and methane. The methane can
be separated, yielding high quality fuel gas.

If all of the municipal solid wastes from the major urban
communities in the Unites States were converted by this process,
we could produce several percent of the total U.S. consumption of
natural gas. However, the value of the energy from this process
will be only a fraction of the cost of the operation. Most of that

cost will have to be offset by the value of the metals and other materials derived as part of the conversion.

Anaerobic digestion is also suitable for conversion of manures from feed lots and other farm operations. Many anaerobic digesters are now being used in China, India, and other developing nations. The Department of Energy has been supporting the development and demonstration of manure anaerobic digesters for use in feed lots in the United States. These can have a significant benefit of reducing the sever water pollution normally caused by such operations.

Gasification of Wood Wastes

Much of the biomass in a tree harvested for lumber is unused, either left in the woods or lost at the sawmill as scrap or sawdust. Some of this waste material is burned by the lumber and paper industries for production of heat and electricity for their processes. However, because of the size and ready availability of this resource, there is significant interest in other ways of using wood wastes. One of the approaches now being developed and tested by a number of firms is air gasification. In this process, chipped wood wastes are combined with a minimal quantity of air and steam to yield a pyrolysis char and a low energy content gas. The gases produced are primarily carbon monoxide, hydrogen, and nitrogen. It should be noted that biomass generally is easier to gasify than is coal. This is because of the better ratio of hydrogen to carbon in the biomass feedstock. There exists a great variety of designs for biomass gasifiers, each with certain advantages (Reed, 1979).

The crucial parameter determining the mix of products from biomass gasification is the amount of oxygen present in the gasifier. The useful parameter is the equivalence ratio: the ratio of oxygen present to the amount of oxygen that would be needed to convert all of the biomass to carbon dioxide and water. At very low equivalence ratios, the primary products of gasification at equilibrium are hydrogen and carbon monoxide, with small amounts of carbon dioxide and water. As the equivalence ratio is increased, the percentage of hydrogen in the product decreases, and the amount of carbon monoxide at first increases. At an equivalence ratio of about 20%, the carbon monoxide production is at a maximum, and the percentage of water and carbon dioxide reaches a minimum. At higher equivalence ratios, the amounts of hydrogen and carbon monoxide decrease, and the amounts of carbon dioxide and water increase. Thus, most gasifiers are operated at fairly low equivalence ratios, but the exact amount of oxygen depends upon the desired product mix.

Just as important as the equilibrium composition indicated by the oxygen content in the gasifier, are the dynamic chemical and physical processes that occur in biomass gasification. There are

many different designs of gasifiers involving updrafts or downdrafts
or various types of mixing. All of these designs are intended to
achieve some desirable mechanical and chemical interaction of the
biomass with hot gases which leads to a maximum production of the
desired components and a minimum of the undesirable components,
especially tars and solid materials.

A number of large gasifiers are now being used in the forestry
industries in the United States. These produce a good quality, low
energy content gas which is used for on-site energy requirements.
Many other types and sizes of gasifiers have been operated. One of
the more interesting is the small gasifier that mounts on the back
of a car and provides a gas for operation of the engine. Such small
automobile gasifiers were used in Europe during World War II, and
they now enjoy a new popularity.

Rapid Pyrolysis of Biomass to Gasoline

I now discuss one example of the many advanced techniques for
conversion of biomass that are now being developed. These advanced
processes offer potential advantages of lower cost, higher efficiency,
or more desirable products, than given by the traditional conversion
techniques. Researchers in China Lake, California, have been
exploring an interesting and unusual process of very rapid pyrolysis
of biomass for production of gasoline or gasoline-like compounds
(Diebold, 1978). The process involves the very rapid heating of
finely divided biomass, resulting in high yield of a synthetic gas.
Subsequent treatment of this gas produces good yields of olefins
which are valuable precursors to gasoline. Bench scale experiments
of this process have been underway. One of the promising variants
of this process would use concentrated solar radiation to provide
the required rapid heating of the biomass. This heat input by solar
radiation could increase the net production of liquids from the
biomass feedstock.

The first step in this process is preparation of the biomass
by grinding it to a fine, dry powder. It is then mixed with steam
and carbon dioxide and blown through a high temperature furnace,
maintained at about 800°C. Rapid pyrolysis occurs. The gases from
this step are then compressed to about 50 atm pressure and enter a
processor for polymerization to gasoline. The yield of liquid fuels
from this process is only a fraction of the feedstock in the experiments conducted thus far. However, process imporvements are hoped
for.

From sources of biomass in agriculture, forestry, and municipal wastes, and with the conversion techniques I have discussed
above, most developed nations could derive about 10% of their total
energy requirements. However, major policy issues will arise to
restrict the development of this biomass resource and affect the way

in which it is used. In developing nations, much higher percentages of energy requirements could be derived from biomass, but
there will arise issues of the renewable nature of biomass production. The future may see severe policy conflicts in these nations
between energy production and the protection of their natural ecosystems.

Hydrocarbon species

As an alternative to the kind of biomass production and conversion techniques described above, some researchers are investi-
gatig plant species that directly produce valuable hydrocarbons
(Benedict, 1979). It has been long known that many plants produce
hydrocarbons as part of their natural growth. The most familiar
of these is the rubber plant. However, several other species have
been grown in the United States for production of hydrocarbons.
A brief effort was made during World War II to grow guayule in the
southwestern United States as a substitute for the natural rubber
supplies unavailable during that period. However, there has been
little commercial development of hydrocarbon species since that
time.

A large number of hydrocarbon-producing species have now been
identified. These include guayule, rabbit bush, goldenrod, creosote
bush, sassafras, euphorbia, milkweed, and jojoba. These plants
contain a number of potentially valuable hydrocarbons, including
waxes, fatty acids, resins, glycerol, tanins, and hydrocarbon
polymers. Techniques have been developed for assaying the content
of these species and measuring the presence of these useful hydrocarbons.

At present, production and demonstration plantings have been
made only of jojoba and guayule. Jojoba serves as a whale oil
substitute and guayule as a substitute for natural rubber. Methods
for studying the hydrocarbon composition of these plants are being
improved. Researchers are screening several hundred plant species
to find those that show high rates of production of desirable hydrocarbons. Some of these researchers now estimate that annual production of hydrocarbons by such species can be as high as eight
barrels of oil per acre per year (about 1 metric ton of oil per
acre). Some argue that this production rate can be maintained only
on good quality soils with high water inputs (i.e., in direct competition with agriculture). However, other researchers are optimisti
that this production rate can be obtained in poor quality desert
soils without additional irrigation. If true, this would mean that
the deserts of the southwest United States could become a significant
source of hydrocarbons. However, it should be noted that the hydrocarbons derived from such plants will probably be more valuable as
petrochemical substitutes than as fuels. Thus, their impact upon
our energy situation might be small.

Ocean Kelp Farming

One way to avoid a conflict between agriculture and biomass energy production is to escape to the open ocean. As mentioned in the first lecture, the natural productivity of ecosystems in the open ocean is very low. There is adequate water and sunlight, but the nutrients crucial for plant growth are lacking in surface waters. It has been suggested that we could bring up to the surface nutrients from the deep ocean waters, thus making areas of high biomass productivity. If this proves feasible, then massive biomass energy programs could be undertaken without adverse impact upon land based agriculture and natural ecosystems. Experiments on such an approach are now underway off the coast of California (Bryce, 1978).

The biomass being produced in these experiments is a seaweed, giant brown kelp. This seaweed naturally occurs in shallow waters along the California coast and has been harvested in the past as a source of iodine. It is a very fast growing seaweed when conditions are right. It requires support for its roots, good sunlight, and adequate nutrients. All but the last can easily be provided at the surface in the open ocean, but special means will be necessary to bring nutrients to the seaweed.

The conversion process being considered for the kelp is anaerobic digestion. This process would yield methane and carbon dioxide. The approach is considered by the gas industry in the United States as being potentially a major renewable source of natural gas.

A test farm based upon this concept is now operating. The mechanical structure of the test farm includes a long pipe and pump which brings up nutrient rich waters from 2,000 feet below the surface, and a series of struts and ropes to which the kelp fasten to support their growth. Divers isolate small kelp plants from their natural habitat in shallow waters and transplant them to the ropes on the test farm structure. The nutrient rich water from the deep ocean is released from a structure of pipe around the kelp, thus increasing significantly their rate of growth. Results of this work are encouraging thus far, but it is much too early to evaluate the commercial feasibility of the concept.

Options for Photochemical Conversion

The most advanced and the most speculative of all solar energy technologies is the direct photochemical conversion of sunlight to fuels. While no practical techniques for this conversion have yet been demonstrated, I consider it to be the most promising approach for the long term development of solar energy as a major energy source. We can distinguish four basic routes possible for photochemical conversion (Hall, 1979; Porter, 1976): first, biological conversion by a natural organism, usually bacteria or algae; second, in vitro conversion by systems that incorporate elements of natural photosynthetic systems; third, photo chemical conversion by synthetic chemical systems; and fourth, photoelectrochemical conversion, which is a hybrid of photochemical and photovoltaic conversion technologies. I will describe each of these briefly.

A number of microorganisms are known to naturally produce hydrogen, though under somewhat unusual circumstances. These micro-organisms include photosynthetic bacteria, green algae, and cyanobacteria. They are observed to produce hydrogen from water and sunlight in conditions in which they are protected from free oxygen. We now understand something of the chemical processes that occur in these organisms that lead to the production of hydrogen (Weaver, 1979). These processes are usually a complex arrangement of individual reactions. Several photons of light are absorbed in the conversion process, and a large number of extraneous reactions and compounds are involved. As a result, the overall efficiency of hydrogen production is very low. This low efficiency probably limits the ultimate usefulness of this approach to photochemical conversion.

To obtain efficiencies higher than those seen in the natural organisms, we can take the approach of developing in vitro conversion systems. In such a system, we would isolate the natural photosynthetic reaction centers from organisms, either plants or microorganisms, and arrange these in an engineered system to produce hydrogen. We can then avoid side reactions and direct all the energy collected toward the desired product. A problem in this arrangement is that we must protect the hydrogen-producing enzymes from free oxygen. They are rendered inactive by even small concentrations of free oxygen. Another problem is these reaction centers are very unstable: they have a very short productive life when isolated from the natural organism. We have only vague ideas as to how to stabilize these reaction centers and maintain their production of hydrogen in vitro.

We avoid the instability of natural photosynthetic systems by developing our own completely synthetic chemical system for photochemical conversion. Many photochemical reactions are known which could be coupled to hydrogen production. However, most of these reactions can be driven only by ultraviolet light. We require a system which can operate efficiently from the visible light of the solar spectrum. No appropriate photochemical system for this conversion is now known, though research on a number of promising ideas is continuing.

Another approach involves combining a semiconductor material for photovoltaic conversion with various liquid electrolytes. In

such a system, the free charges created by the photovoltaic effect
are captured and converted into chemical energy. A number of such
systems have been operated, though again at rather low efficiencies.
Research in this area is also underway in a number of laboratories.

Thus, we see that there are a number of promising approaches
to photochemical conversion being actively investigated. None of
these has advanced beyond the stage of laboratory research. However, this work is of great importance because of the very desirable
features possible in photochemical conversion. Theoretically, conversion efficiencies could be very high, up to 20%. This should be
contrasted with the normal efficiency of plant growth, about 0.5%.
With the high efficiency theoretically possible from photochemical
conversion, a modern nation could derive all of its energy needs
from a small fraction of its land area. Further, this could be
done without significant adverse impact upon the operation of its
agricultural systems or its natural ecosystems. The energy is
produced in a naturally storable form, thus avoiding the complex
issues of interfacing variable wind or solar energy sources to the
patterns of normal energy demands. Thus, I expect photochemical
conversion to be of great importance in the ultimate development of
a renewable energy society.

RENEWABLE ENERGY POLICY

In the previous lectures, we have reviewed the current state-
of-the-art and the directions being taken in research and development in the various solar energy technologies. But solar energy
involves more than just technical issues. It has policy aspects
which include complex environmental, social, and political issues.
In this last lecture, I will cover a number of these issues and
offer some of my own thinking on how we should approach them.

In a market economy, which in many respects the United States
still is, the core policy questions can be stated: are private firms
and consumers investing in renewable energy technologies at the
socially optimum rate? If not, what policies can be effective in
correcting the level of investment in renewable energy (Schiffel,
1978)? I think these two questions capture the important issues
surrounding the development and use of solar energy. The first
includes all questions of private business and consumer behavior
as contrasted with behavior that gains social benefits enjoyed by
the nation as a whole but not accruing to individuals in the market
system. The second brings in the nasty but crucial issue of whether
any government policy can be effective in correcting what might be
an adverse outcome of the market system. Thus, this question
expresses skepticism about the ability of government to actually
make things better.

Many would not agree with the formulation of the issues as I have given them. To them, the worst possible fate that might befall solar energy is for solar technologies to be developed and marketed by the same major corporations that are seen to be responsible for many of our current social, political, and environmental ills. To these people, the development of solar energy offers a chance to make some fundamental revisions in the structure of society. While they might press for solar energy for these reasons, I see no strong connection between their political and social goals and solar energy. Those of us who find other, less political, virtues in the use of renewable energy can support its development while maintaining our own views and ways of expressing our position on social and political issues.

In this last lecture, I will not be able to address all of the important policy issues that are being debated actively in the U.S. and elsewhere concerning the development of solar energy. I will cover four areas that are central to the debate. The first has to do with the environmental costs and benefits of solar energy in comparison with conventional energy technologies. The second is the issue of the employment effects of a transition to solar energy by the evolution of industries and businesses to provide solar energy hardware and services. Third is the potential role of solar energy in aiding the current plight of the developing nations who face high imported oil costs that endanger their economic development. Finally, I will discuss various theoretical structures for thinking about the long term transition to a society based upon the use of renewable energy resources.

Environmental Costs and Benefits

Many of the issues surrounding the relative value to society of solar energy and conventional energy forms concern the externalities of these energy sources. An externality is any cost or benefit which does not accure to the person responsible for the activity. The classic examples are environmental costs. For example, the simple smokestack of an industry allows the business to get rid of pollutants at low cost and yet subjects the surrounding residents to adverse health effects and economic costs of air pollution. There is an externality here because the business does not pay all of the costs imposed by its operation. The common understanding is that conventional energy technologies have significant externalities that should be considered in their assessment. Certainly we have made progress in recent years towards forcing an internalization of many of the environmental costs of conventional energy sources. For example, the cost of installing and operating a stack-gas scrubber on a coal power plant is an internalization of what used to be an externality. Yet, there are still many environmental costs associated with conventional energy sources. We will never be

able to internalize all of these costs. They will continue to be
unpaid by those responsible for them and will continue to be an
inducement to develop public policies to alter the operation of
the normal market process in selection of energy forms.

It is sometimes claimed that solar energy technologies are
free of external environmental costs and thus deserve a significant
subsidy relative to the conventional sources with their unpaid
external environmental costs. However, a moment's thought shows
that solar technologies are not entirely free from external environmental effects. One must, however, in the case of solar technologies,
consider the full life-cycle of the solar facility in evaluating its
environmental effects. Part of the social cost of the solar facility
is the environmental cost of emissions from the factories that make
the solar hardware. Some of these solar technologies, for example,
biomass gasification plants, will have significant environmental
costs during their period of operation as well. Finally, there will
be some environmental costs associated with removing, dismantling,
and recycling the materials of the solar hardware after the useful
life of the system.

In analyses performed at SERI, we have compared the total life-
cycle emissions from various solar technologies with the corresponding total life-cycle emissions of a coal electric power plant
that produces the same energy output (Lawrence, 1979). We assumed
modern environmental control standards for the coal facility and
either the current state-of-the-art or a direct extrapolation of
present capabilities for solar technologies. Some extrapolation
from present technology is necessary to develop a reasonable base
for assessment of many solar technologies. However, an attempt
was made to provide a conservative estimate of the total emissions
due to the construction, operation, and dismantling of a solar
facility. Figure 8 shows the results obtained for one of the pollutants considered, particulates. It can be seen that the worst
offender among these technologies is not the coal electric plant,
but the common wood stove. Almost equal to the coal electric plant
in particulate emissions is a biomass steam electric plant. But
generally, there are major savings in pollutant emissions of the
solar technologies in comparison to the coal electric plant. This
result for particulates generally holds true for all the pollutants
and for the estimated impacts on human health as well (Yokell, 1979).
We find major savings for the solar technologies in comparison with
conventional technologies with regard to life-cycle environmental
effects. This result is not supported by one analysis which has
acquired a great deal of publicity recently (Inhaber, 1979;
Holdren, 1979). However, I find that the works by Lawrence and
Holdren a more objective and more reliable indicator of likely
effects, and I agree with their general conclusions as to the
environmental benefits of solar energy.

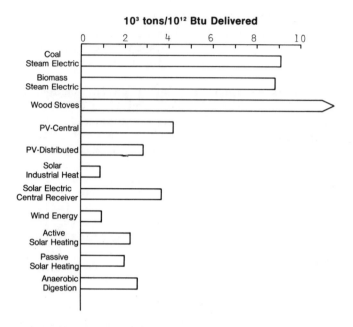

Figure 8. Total particulates released per unit of energy delivered
over the life-cycle of coal-electric and solar energy
technologies (Lawrence, 1979).

        The analyses just discussed look at only one facility at a
time and make direct life-cycle comparisons between types of
facilities. However, our major interest is not in the effect of
a single facility, but in the total national impact of increased
use of solar technologies on a wide scale. Thus, we must sum the
effects of many individual facilities to produce an estimate of
national total pollutant releases. To do this, you need a scenario
of the rate of solar technology deployment. In analyses done at
SERI, we have analyzed a scenario that corresponds to attainment
of President Carter's goal of 20% solar energy in the United States
by the year 2000. We have used two different national energy-
environmental models for this work. One is the large model
developed by the Brookhaven National Laboratory; the other is a
complex model developed by the Environmental Protection Agency
called the Strategic Environmental Assessment System (SEAS). We
have used these models to analyze two scenarios (Yokell, 1979).
One is a base case in which the use of conventional energy sources
continue to be dominant, and there is little use of solar energy
by the year 2000. The other scenario we call the High Solar Scenario
and it corresponds to an accelerated schedule of development of
solar energy and attainment of the 20% goal by the year 2000.

Table 6.   Effects of Accelerated Use of
Solar Energy on National Pollutant Releases
in the Year 2000

| | BNL Model | | SEAS Model | |
Pollutant (10^6 tons)	Base Case	High Solar	Base Case	High Solar
Particulates	11.2	9.3(-)	10.2	10.8(+)
Sulfur Oxides	11.5	8.5(-)	26.3	23.2(-)
N Oxides	26.8	22.5(-)	23.3	21.1(-)
Hydrocarbons	5.7	5.5(-)	10.0	9.7(-)
Carbon Monoxide	15.6	15.2(-)	59.9	60.1(+)
Suspended Solids	0.2	0.1(-)	1.9	2.7(+)

Table 6 shows the total national pollutant releases in the year
2000 in these two cases as indicated by the two different models.
Several things can be noted. First, the differences between the
two models for the same scenario are generally larger than the differences between the two scenarios as predicted by the same model.
Thus, the uncertainties in our projection techniques are very large
compared to the effects we are analyzing. Second, there is no clear
and unambiguous reduction in the emissions in the High Solar scenario
as analyzed by the SEAS model. With the BNL model, there are reductions in all of the pollutants, but the changes are very small and
are certainly within the likely uncertainties of the model.

These models lead to a conclusion that seems contrary to the
comparison of life-cycle effects I discussed above. The reason for
this difference is simple. In a scenario that involves rapid transition to solar energy, there is a very high rate of production of
solar hardware during the first few decades. If you recall that much
of the pollutant emission during the life-cycle of a solar facility
comes during its production and installation, then you can realize
that in the next 20 years, the High Solar scenario has high emissions
exactly because a large number of solar facilities are being constructed. If we were to run these models further into the future,
we would then start to see that the High Solar scenario has significantly reduced pollutant emissions relative to the Base Case scenario.
However, because the environmental impacts of solar technologies come
primarily at the beginning of their life-cycle, no environmental
benefits are seen in the early stages of a rapid transition to solar
energy. Thus, we have a paradox of sorts. Solar has great

environmental benefits, but we cannot gain those benefits immediately. To attempt to grasp them rapidly is to cause those benefits
to slide forward into the 21st century.

Employment in Solar Energy

It is claimed that one of the benefits of a rapid development
of solar energy is the creation of new employment opportunities.
It is argued by some that solar energy can help solve the problems
of unemployment that seem characteristic of modern industrial
society. However, no study to date offers a comprehensive or pro-
fessionally defensible analysis of solar energy employment. Some
of the studies done to date have been politically motivated, and
their methods and data are suspect. Some have failed to recognize
the complexity of the issues involved in this area. All have
suffered to some degree from unavoidable deficiencies in the current
state-of-the-art of theory, data, and methods for analysis of macroeconomic and employment issues (Mason, 1978).

In a recent study by SERI, the Brookhaven National Laboratory,
and Dale Jorgenson Associates, we have attempted to assess the
impacts of solar energy under a variety of assumptions (Mason, 1979).
Most importantly, we have considered the two cases of solar energy
being cost competitive with conventional energy technologies, and
not cost competitive with those conventional energy sources. In
the latter case, the introduction of solar energy might be brought
about either by government policy through incentives or mandate,
or by willingness of consumers to pay a premium for what they perceive to be special benefits of solar energy. This study suffers,
as do all others done in this field, from the inadequacies of
present theory, data, and methods in conducting macroeconomic analysis. However, within the accuracy now possible, our conclusions are
these: First, if solar energy is cost competitive with the conventional energy technologies it displaces, we find no significant
impact on total employment. There are only a few shifts of employment between various sectors. This is true even in the case that
the deployment of solar energy is rapid enough to meet President
Carter's goal of 20% solar energy in the year 2000. On the other
hand, if solar energy is not cost competitive with the energy technologies it displaces, we then find a slight negative impact on
total employment in the economy. This consequence is a result of
the movement of investment capital from other investments in the
economy into an investment in solar energy systems. Thus, this conclusion depends upon the operation of a macroeconomic model which
attempts to estimate the effects of investment patterns upon levels
of economic activity and thus upon total employment. This conclusion
must be considered very weak because of the poor state-of-the-art
in macroeconomic modeling.

To see why it is so hard to estimate the employment impacts
of solar energy, let us trace the chain of logic necessary
(Hudson, 1979). We start with the question as to whether solar
energy will require more or fewer economic inputs than the conventional technologies it displaces. Let us, for the moment, assume
it requires net additional economic inputs. Then we must ask the
question, does there exist excess capacity in the economy for the
production of these additional inputs required for solar energy
relative to conventional energy technologies? If there is indeed
such excess capacity, then there is no dynamic macroeconomic impact
of this activity, and there is a direct employment benefit of
solar energy. However, if the answer to this last question is "no,
there is no excess capacity for the inputs required," then labor
and investment must be diverted from other uses in the economy in
order to fabricate the solar energy systems. This then causes a
chain of adverse effects in the economy with the probable results
of lower productivity, slower capital growth, and slower economic
growth. As a result, there may be a negative total impact upon
employment. A similar chain of reasoning can be followed taking
the assumption that solar energy will require fewer economic inputs
than the conventional energy technologies it replaces. If there is
excess capacity in the economy for the input requirements that are
released, then there is no dynamic effect, and only a small decrease
in employment. On the other hand, if there is not any excess
capacity for these inputs, then the reduction in input requirements
for solar energy relative to conventional sources allows labor and
investment to be made available for other uses in the economy. This
will then generally lead to higher productivity, faster capital
growth, and faster economic growth. Thus, we see there is no unique
answer to the question of the employment impacts of solar energy
at the national level. It depends upon the relative requirements
of solar energy for labor and other economic inputs, and depends
upon the current situation of the economy with regard to the availability of those inputs and their uses in other areas of economic
activity.

From all of this, my conclusion is that the employment impacts
of solar energy on the national scale will be fairly small. Indeed,
they will be small enough that we cannot predict them because of
our inability to understand the processes that occur in the national
economy. If one wishes to see the important employment effects of
solar energy, you have to look at the microeconomic level, and see
exactly what sort of skills will be required for solar energy, in
which locales, and compare that to the local availability of that
skill. At this micro level, you will see some positive employment
effects of solar energy, but not effects significant enough to
motivate policies at the national level. In my opinion, national
policy must be based upon other benefits of solar energy than
employment.

Solar Energy in International Development

Solar energy promises to become an important element of the international program of assistance to developing nations. Payment of high fuel prices by non-OPEC developing nations has jeopardized both future industrialization plans and the gains in quality of life made during the preceding two decades. The economies of these nations cannot long support the ever increasing economic penalty of importing OPEC oil. Among the solutions being tried is the rapid deployment of appropriate renewable energy technology, especially in rural areas of these developing nations. However, it is recognized that these nations need assistance both in gaining technical expertise and in funding of solar energy programs.

A variety of programs for international assistance in renewable energy technology are now being undertaken (Ashworth, 1979). These include multilateral donors, bylateral donors, and nongovernmental organizations. Among the multilateral donors, the most prominent are the U.N. Development Programme and the World Bank. Among the bilateral donors, the largest aid programs are those being conducted by the U.S., Canada, and major European nations. Among the nongovernmental orgainzations, prominent activities have been conducted by the Rockefeller Foundation and voluntary agencies of the OECD countries.

Altogether, in excess of $200M a year is now being spent for international assistance projects in renewable energy and appropriate technologies. Eighty-six percent of this, by the estimates made by SERI, come from four nations: U.S., Sweden, Germany, and Canada. The projects include both traditional activities of international assistance that are now being directed toward renewable energy, and also some new initiatives that are special to solar energy. Among the traditional activities are these: institution building, the development of institutions for research or dissemination of information within the nation for renewable energy; manpower training, to train technicians and others to construct and maintain renewable energy systems; and resource creation and management, activities devoted primarily towards the management of ecosystems resources (i.e., biomass). In new initiatives, the following activities are underway: base line data collection, the development of information on energy requirements in rural areas and data on local energy resources; demonstration and field testing, demonstration and evaluation of technologies in the actual conditions in which they must operate (often coupled with training); technology and hardware development, the development of new technologies and hardware to meet the special requirements of these applications or to take advantage of special local resources and expertise; and creation of energy self-sufficient villages, demonstrations of complete economies for rural areas which can operate with self-independence in energy.

While solar energy will not be a magic solution to all the problems of developing nations, it does now appear that there are many circumstances in which solar energy and other appropriate technologies can significantly improve the quality of life in these nations. Efforts in this direction deserve to receive, and are now receiving, the active support of the developed nations.

Theory of Renewable Resources

As the concluding topic in these lectures, I will review various approaches to a theoretical basis for policy of renewable energy resources. The discipline of economics provides most of the terms and methods used in this area of inquiry. However, the concepts involved extend beyond those normally included in economic analysis. Ethical and moral concepts must be dealt with, either explicitly or implicitly. The question at issue can be stated this way: How should society schedule its consumption of finite resources and the adoption of renewable resource technology? The appearance of the word "should" in this question shows that we are working in the area of normative economics; that is, we are dealing with questions of what we consider to be the best choices for society according to some set of values. The ways that one assigns values to resources and their uses determines the theoretical structure and the assumptions one uses in this area of analysis.

Two major schools of thought delineate the broad range of opinions in this area. The position of the neoclassical economists is to let the market decide. That is, they make the presumption in this area, as in other questions concerning allocation of resources, that market forces will provide the optimum scheduling of a transition from a depleted resource to a new resource. At great variance with the position of the neoclassical economists is the school of thought which argues for creation of a steady-state economy. Here it is usually argued that normal market forces will fail to provide an adequate transition to a society which operates efficiently in a situation of limited resources. The gulf between these two viewpoints could hardly be wider.

I personally find much of value in both viewpoints. Each school of thought seems to have important things to say, but each analysis is valid only within a limited sphere. The question I find most provocative and interesting is how one can develop a theoretical structure that bridges these two viewpoints and deals with the issue of the optimum transition from a market economy in a period of cheap resources, to the steady state economy which has achieved a careful balance within a limited world. I know of no adequate bridging structure between these two schools of thought. I will suggest in this part of my lecture some of my personal views on how such a structure might be developed, but I'm only indicating promising ideas for future exploration.

In the literature of the neoclassical school, there is no problem of the exhaustion of finite resources that requires special social policies. The fundamental assumption is that market price gives an adequate signal to consumers and producers of any impending scarcity. Indeed, the viewpoint of this school is that government actions to restrict the operation of market forces will more often disguise and prevent adequate response to the valid signals of scarce resources. The example often given is regulation of the price of natural gas, which seems to have discouraged exploration for new fields and failed to give consumers an adequate price signal of the declining availability of natural gas.

An important part of the argument of the neoclassical economist is that increased effort for production of the resource will be stimulated by price increases. This will cause increased discoveries and availability. Thus, price will increase gradually over time, and resource availability will be assured. Many economists consider that this process can occur without limitation, that is, they do not recognize explicitly in their analyses any real physical limits to resource size or producibility. When they do recognize physical limits to resources, they then argue that the price mechanism will lead to economically more plentiful resources being substituted for less plentiful ones. However, there is little treatment in the literature of how this substitution process takes place, nor much examination of whether the transition occurs in an optimum manner.

Part of the school of thought represented here has focused on the important consequences of technological progress. It has been argued that technological progress will make it possible to keep producing ever larger volumes of extracted goods at declining real marginal costs. Thus, the view here is that technology is a much more important effect than is resource depletion. The evidence for this is said to be a declining real cost over time of goods produced from basic resources. There is some evidence in the history of long term price trends for basic resources that supports this view. In a recent review of this issue, there was found to be ambiguous evidence in most areas (Smith, 1979). It does not seem possible to conclude firmly whether there has or has not been an increase in the real price of commodities due to declining natural resources. Even in the case of petroleum, there does not seem to be clear evidence that would support the idea that the cost of petroleum has increased due to a decline in remaining natural resources. The price increases of the past years are due to accidents of ownership and location of resources in relation to users, and do not indicate any exhaustion of the resource that is leading to increases in actual production cost.

The optimism of the neoclassical school of thought is in stark contrast to the views of the steady-state economists. These two schools differ in the starting point for their analysis. The neoclassical economist starts with the functioning of the existing market system and the historical behavior of prices and production rates. In contrast, the steady-state economist starts with the physical nature of the resource and the geological and ecological limitations to its development. The neoclassical economist assumes technological progress will continue and will be the dominant effect. The steady-state economist more often ignores completely any possibility of technical progress and assumes, perhaps implicitly, that the steady state economy has no technical advances possible within it. These two schools of thought also differ in how they approach economic analysis. The neoclassical economist is primarily positive; that is, he attempts to analyze what is occurring and project what is most likely to occur in the future. On the other hand, the steady-state economist takes a normative viewpoint: he asks not what is, but what should be? Thus, the steady-state viewpoint has an ethical content that is explicit, whereas the ethical content is only implicit in the neoclassical analyses.

The school of thought of the steady-state economist usually is built around the image of a society which is believed to represent a desirable or necessary situation (Daly, 1977). This viewpoint does not start with a description of the current state of society, but rather an ideal society: the steady-state economy, which is best defined as being one in which there is no net consumption of any finite resource. This definition does not actually require that the economy operate in a steady state. However, the term "steady-state economy" has been used frequently in the literature, and I will continue to use it even though it is somewhat inaccurate.

As an obvious conclusion of the requirement for no net consumption of any resource, all nonfuel materials must be recycled; thus copper, silver, aluminum, and other metals must be 100% recycled. However, as we know from the second law of thermodynamics, energy cannot be recycled. Thus, for this society to fit the definition given above, it must derive its only energy resources from solar energy, either directly or indirectly. It is probably useful to relax this last requirement somewhat and include in our thinking those finite resources that are so extensive that they would last long by historical times even if used at rates corresponding to the full energy requirements of a modern society. Thus, I would include both the breeder reactor and fusion energy in this category. In almost everything I say in this last part of the lecture, the term solar energy is interchangeable with these other very large, essentially infinite, energy resources.

The steady-state economy has some other important characteristics. Population must be held constant. The capital stock as measured by physical quantities must be held constant. Steady-state economists also argue that society must maintain a careful balance with the natural ecosystems of the world. To do this, the physical

thoughput of materials and energy in the economy must be minimized
by selection of high efficiency means of providing amenities. One
consequence of this is that Gross National Product (GNP) is no
longer a significant measure of economic well being. Indeed, it
can be said that one objective of a steady-state economy is to
minimize GNP for a given level of social welfare. However, this
last rule is only useful for GNP as measured in terms of the flow
of physical quantities. With other measures, GNP can grow because
of the potential for technological progress in a steady-state
economy.

The steady-state economists generally ignore the possibility
of further technological progress: their analyses do not include
it explicitly. However, I think this is an important and fundamental
error. I think it quite possible, and even most likely, that a
society that fits the above definitions of no net consumption of
finite resources, will have significant economic progress through
increasing sophistication of its technology. Thus, improvements
in technology would allow yet better benefits for its population
to be derived from its use of completely recycled finite resources
and its use of renewable energy resources. In a measure of economic
activity that represented a proper extension of Gross National
Product, this would mean that economic productivity would continue
to increase in a steady-state society even though the net use of
physical resources does not increase.

It should be clear from the above summaries that there is
little in common between the views of neoclassical economists and
steady-state economists. I personally believe the neoclassical
economist offers a useful, richly detailed, and pragmatic description
of the current state of society, while the steady-state economist
provides a useful, soundly based, and optimistic description of
the long term future of humanity. The question I turn to next is
how we can provide a bridge between these two viewpoints and thus
start to understand how society might make the transition from our
present economic system to one that is indefinitely sustainable.

In this concluding part of my lectures, I cannot offer any
firm answers. All I can do is suggest some questions and some
assumptions I think useful for future thinking about these issues.
This viewpoint is highly personal. I offer here my own ideas as
to how one should proceed in this analysis, and I am sure that
many would disagree with one or more of these points. As we enter
this discussion, you will see that many of the questions turn upon
values and disagreements as to what the future of humanity should
be like.

First, and as a fundamental point, I suggest that in this
analysis, benefits to future generations should not be discounted.
That is, the risk-free discount rate in this analysis should be set

to zero. Some would respond to this suggestion by saying that the
discount rate represents an inevitable cost of deferring the use
of money. In that view, the discount rate represents a natural part
of an economic process, and we cannot ignore it. However, I would
argue that discount rates used for short term analyses (i.e.,
comparing an investment now with one ten years from now) are not
useful for analyses that compare actions centuries apart. Indeed,
the straightforward application of an annual rate of discount to
processes that occur over hundreds of years leads to ludicrous conclusions. For example, if one discounts the present worth of the
entire North American Continent back to the time of Columbus' trip,
one finds that, to Columbus, the present value of the United States
could not justify the cost of his trip. Clearly, he should have
stayed home. The fundamental question here is whether in an analysis
of benefits to human beings occurring at greatly different times,
should we discriminate systematically against one generation in favor
of another? I argue we should not so discriminate. Thus, our
discount rate should be zero.

I think it necessary that our analyses include explicitly
physical constraints to resource use. Thus, I would not allow
econometric models in which the availability of a resource depends
only upon prices and in which a certain schedule of prices can lead
to ever increasing levels of production. Further, we must treat in
some way the nature of geological resources: as resources that are
easy to withdraw are exhausted, the remaining resources will require
larger investments in energy, capital, and labor to withdraw and
utilize. It is, of course, extremely difficult to estimate the
physical extent of resources not yet carefully explored. However,
I think it far better for our purposes to include a guess as to the
extent of a physical resource than it is to deny the existence of
any physical constraint. Hopefully, the major conclusions of our
analysis with regard to appropriate policies and strategies will
not be highly sensitive to the exact size of resources.

If the only resources available to society are the finite
resources discussed above, then our analysis must lead to a conclusion of ultimate disaster. We are led, therefore, to the
requirement that our analysis must explicitly include the possible
use of renewable resources as alternatives to finite resources.
However, to be an accurate picture of the transition we face, it
should also indicate that there are significant costs to renewable
energy technologies, and that at the beginning point of this transition, the renewable technologies will be much more expensive than
the finite resources. Our analysis must also include the fact that
the development and use of renewable technology will always require
investments of energy, capital, and labor. Thus, an important
question for the long term future is how efficiently it is possible
to use these inputs in order to derive renewable energy resources
in useful forms.

In a renewable resource future, it is important that the level of the capital stock be explicitly included. This is because benefits to the population in such an economy are provided largely by the amount of the capital stock, rather than by the flow of commodities. Important questions for the analysis are how the capital stock should grow over time during the transition from finite resources to renewable resources, as well as what is the optimum level of capital stock in the steady-state situation? I expect that one of the conclusions of this part of the analysis will be that the rate of investment and saving in society is crucial to a successful transition to renewable energy. This puts the policy question at the level of the individual and his tradeoff between immediate consumption and savings. The present very low rate of personal savings in the United States is not an encouraging indicator of our readiness to proceed with the kind of transition being discussed here.

Finally, two formal points about how we should define the optimum schedule of the transition. As in normative economics, we assume that the distribution over time of benefits should be a Pareto optimum. That is, it should be impossible to alter this distribution of benefits in any way so as to further increase the benefits to one generation without reducing the benefits to another generation (Herfindahl, 1974). From this rule, we conclude that the optimum scheduling is one in which we do as well as possible for each generation up to the point at which harm is imposed on another generation. However, this rule will often not lead to a unique scheduling of the transition. Another condition may be necessary to make the schedule unique. Here I would adopt the principle used by Rawls in his development of the theory of justice (Rawls, 1971). Thus, we would attempt to make the distribution of benefits as uniform as possible, so that if people had a free choice of which generation to join, they would be indifferent as to that choice.

It may be worthwhile to look briefly at a simple model of the use of an exhaustable resource in order to see the kind of analytic structure that might be useful in pursuing the directions I have indicated above. To do this, we look at the simple model used by Koopmans to illustrate the basic issues in the use of an exhaustable and nonsubstitutable resource (Koopmans, 1973). Koopmans presents the following model. One wishes to maximize

$$U = \int_0^T e^{-rt} u(c(t)) dt$$

subject to

$$\int_0^T c(t) dt = F$$

and

$$c(t) \geq c_{min} > 0 \quad \text{for} \quad 0 \leq t \leq T$$

Here U is the total benefit over time, that is, it is a measure
of total utility. The utility derived by society at an instant
from its consumption of the resource is given by u, and the rate
of consumption by c(t). The second equation gives the condition
that the total consumption of the resource over time has a finite
limit. The next condition indicates that there is a minimum level
of consumption below which society cannot continue. As an inevitable consequence of these two requirements, there is a finite
lifetime to society. That is, the total duration of society in
this model must be less than F/c_{min}. This, then, is a model of
extreme pessimism. Society is living with a finite resource that
inevitably will be exhausted at some time in the future. The only
question is how the few generations that can survive will schedule
the use of this resource before they come to an end. In this model,
there is a discount rate, r, applied to the utility of consumption
of the resource. With large values of r, the solution to the model
given above has a high level of consumption for the early generations,
declining to c_{min} rather quickly with a resulting rapid end to
society. With r set equal to zero, the consumption of the resource
does not change over time, and the optimum level of consumption
depends upon the nature of the utility function, $u(c(t))$.

    This model given above is a starting point from which we can
think about how to develop an analytical structure to model the
transition from finite to renewable resources. However, the model
must be extended in many ways to deal with the issues I have outlined. First, benefits must derive from consumption of finite
resources, from the existing capital stock, and from use of renewable
resources. Second, economic production must be divided between
consumption and the creation of capital stock, including the creation
of stock for the development of renewable resources. At issue is
how much of the finite energy resource should be consumed for
immediate enjoyment, and how much should be used for building solar
energy systems to provide energy in the future? Economic production
should result both from depletion of finite resources and from use
of renewable resources by the capital stock devoted to that purpose.
The problem can then be stated as one of maximizing the integral
from now to the infinite future of the instantaneous utility of
the operation of the economy thus described, with the optimum constrained by the conditions of Pareto and Rawls.

    This model is obviously a highly simplified vision of an
economic society. It does not answer many of the questions of
immediate policy importance. However, it is not intended to replace
normal economic analyses for short term decisions. Rather, it

provides a structure for thinking about the far future and the
kinds of transitions that must eventually occur. In this regard,
it might give us some useful insight into the implications of
our present patterns of behavior toward resources and how that
behavior must eventually be modified.

I think we are all in agreement that renewable resources must
be developed, and we are hopeful that society can succeed in making
a pleasant and smooth transition to their use. In these lectures,
we have seen some of the technologies now being developed which
offer a means of achieving a renewable energy society. It is clear
we have far to go, and yet I hope I have made it clear in these
lectures that our progress to date has been impressive. We should
have high hopes that this progress will be continued to the great
benefit of future generations.

References

B. Anderson and M. Riordan, The Solar Home Book, 1976. Harrisville,
     New Hampshire, USA: Chesire Books.
J. Ashworth, Renewable Energy Sources for the World's Poor, 1979,
     Golden, Colorado: Solar Energy Research Institute report
     SERI/TR-51-195.
L.L. Anderson and D.A. Tillman, Fuels from Waste, 1977. New York,
     NY: Academic Press.
H.M. Benedict and B. Inman, A Review of Current Research on
     Hydrocarbon Production by Plants, 1979. Golden, Colorado:
     Solar Energy Research Institute report SERI/TR-33-129.
P. Berdahl, California Solar Data Manual, 1977. Berkeley,
     California: Lawrence Berkeley Laboratory report LBL-5971.
M.K. Boardman, The Energy Budget in Solar Energy Conversion in
     Ecological and Agricultrual Systems, 1977, in Living Systems
     as Energy Converters, R. Buret editor, Amsterdam: North-
     Holland.
T. Brumleve, Recommendation for the Conceptual Design of the
     Barstow, California Solar Central Receiver Pilot Plant-
     Executive Summary, 1977. Livermore, California: Sandia
     Laboratories report SAND 77-8035.
A.M. Bryce, A review of the Energy from Marine Biomass Program,
     1978. In Energy from Biomass and Wastes, D.L. Klass, Chicago,
     Illinois: Institute for Gas Technology.
P. Call, National Program Plan for Absorber Surfaces R&D, 1979.
     Golden, Colorado: Solar Energy Research Institute report
     SERI/TR-31-103.
D. Costello, Photovoltaic Venture Analysis, 1978. Golden, Colorado:
     Solar Energy Research Institute report SERI/TR-52-040.
D.A. Curto and Z.D. Nikidem, Solar Thermal Repowering, 1978.
     McLean, Virginia: Mitre Corporation report MTR-7861.
H.E. Daly, Steady-State Economics, 1977. San Francisco, California:
     W.H. Freeman and Co.

J. Diebold and G. Smith, Conversion of Trash to Gasoline, 1978.
    China Lake, California: Naval Weapons Center report
    NWC-TP-6022.

J.W. Doane, A Government Role in Solar Thermal Repowering, 1979
    forthcoming. Golden, Colorado: Solar Energy Research
    Institute report SERI/TP-51-340.

DOE, Ocean Systems Program Summary, 1978. Washington, DC: U.S.
    Department of Energy publication DOE/ET-0083.

F.S. Dubin and C.G. Long, Energy Conservation Standards, 1978.
    New York, NY: McGraw-Hill.

H. Ehrenreich, Solar Photovoltaic Energy Conversion, 1979. New
    York, NY: American Physical Society.

P.R. Ehrlich, A.H. Ehrlich, and J.P. Holdren, Ecoscience, 1978.
    San Francisco: W.H. Freeman.

S. Flaim, Soil Fertility and Soil Loss Constraints on Crop Residue
    Removal for Energy Production, 1979. Golden, Colorado: Solar
    Energy Research Institute report SERI/RR-52-324.

H.C. Goddard, Managing Solid Wastes, 1975. New York, NY: Praeger
    Publishers.

E.W. Golding, The Generation of Electricity by Wind Power, 1976.
    London: E&F.N. Spon Ltd.

C.G. Grosskreutz, 1979. Briefing for the U.S. Department of Energy
    Division of Solar Technology. Solar Energy Research Institute,
    Golden, Colorado, April 1979.

D. Hall, Photochemical Conversion of Solar Energy, 1979. In Annual
    Review of Energy, Volume 4. Palo Alto, California: Annual
    Reviews, Inc.

O.C. Herfindahl and A.V. Kneese, Economic Theory of Natural Resources,
    1974. Columbus, Ohio: Charles E. Merril Co.

D.I. Hertzmark, A Preliminary Report on the Agricultural Sector
    Impacts of Obtaining Ethanol from Grain, 1979. Golden, Colorado:
    Solar Energy Research Institute report SERI/RR-51-292.

J.P. Holdren, Risk of Renewable Energy Sources: A Critique of the
    Inahaber Report, 1979. Berkeley, California: University of
    California Energy and Resources Group report ERG-79-3.

E.A. Hudson, Macroeconomic Effects of Solar Energy, 1979. Cambridge,
    Mass.: Dale W. Jorgenson Associates.

R.L. Hulstrom, Insolation Models, Data and Algorithms, 1978. Golden,
    Colorado: Solar Energy Research Institute report SERI/TR-36-110.

H. Inhaber, Risk with Energy from Conventional and Non-conventional
    Sources, 1979. Science, 203: 718-723 (23 February).

C.G. Justus, 1978. Winds and Wind System Performance, Philadelphia:
    Franklin Institute Press.

E. Kahn, Compatibility of Wind and Solar Energy with Conventional
    Energy Systems, 1979. Annual Reviews of Energy, Volume IV,
    Palo Alto, California: Annual Reviews, Inc.

D.L. Klass, Energy from Biomass and Wastes, 1978. Chicago, Illinois:
    Institute for Gas Technology.

T.C. Koopmans, Ways of Looking at Future Economic Growth, Resource
     and Energy Use, 1973. In M.S. Macrakis, ed, Energy, Cambridge,
     Mass.: MIT Press.
J.F. Kreider and F. Kreith, Solar Heating and Cooling, 1977. New
     York, NY: McGraw-Hill.
F. Kreith and J.F. Kreider, Principles of Solar Engineering, 1978.
     New York, N.Y.: McGraw-Hill.
K. Lawrence, The Net Environmental Benefits of Solar Energy
     Technologies, 1979. Golden, Colorado: Solar Energy Research
     Institute report SERI/TP-53-322.
W.D. Marsh, 1979. Requirements Assessment of Wind Power Plants in
     Electric Utility Systems, Volume II, Palo Alto, California:
     Electric Power Research Institute Report ER-278 V.2.
W.D. Marsh, 1979 a.   Requirements Assessment of Wind Power Plants
     in Electric Utility Systems, Volume III, Palo Alto, California:
     Electric Power Research Institute Report ER-978 V.3.
E. Mazria, The Passive Solar Energy Book, 1979. Emmaus, Pennsylvania:
     Rodale Press.
B. Mason, Solar Energy Commercialization and the Labor Market, 1978.
     Golden, Colorado: Solar Energy Research Institute report
     SERI/TP-53-123.
B. Mason, Macroeconomic Impacts of Solar Energy, 1979. Golden,
     Colorado: Solar Energy Research Institute report SERI/TP-321.
J. Mills, Solar Industrial Process Heat Conference Proceedings, 1978.
     Golden, Colorado: Solar Energy Research Institute Report
     SERI/TP-49-065.
G. Porter and M.D. Archer, In Vitro Photosynthesis, 1976.
     Interdisciplinary Science Reviews, Vol. 1, No. 2.
P.C. Putnam, Power from the Wind, 1948. Florence, Kentucky,
     Van Nostrand Reinhold.
J. Rawls, A Theory of Justice, 1971. Cambridge, Mass.: Harvard
     University Press.
T. B. Reed, A Survey of Biomass Gasification, 1979. Golden, Colorado:
     Solar Energy Research Institute report SERI/TR-33-239.
A.H. Rosenfeld, Building Energy Compilation and Analysis, 1979
     forthcoming. Berkeley, California, Lawrence Berkeley Laboratory
     report.
D. Schiffel, Solar Incentives Planning and Development, 1978.
     Golden, Colorado: Solar Energy Research Institute report
     SERI/TR-51-059.
B. Shelpuk, Alternate Cycles Applied to Ocean Thermal Energy
     Conversion, 1979. Golden, Colorado: Solar Energy Research
     Institute report SERI/TP-34-180.
C. Smith, 1979. Briefing for the U.S. Department of Energy Division
     of Solar Technology. Solar Energy Research Institute, Golden,
     Colorado: April 1979.
V.K. Smith, ed. Scarcity and Growth Reconsidered, 1979. Baltimore,
     Maryland: Johns Hopkins University Press.
J. Thornton, Comparative Ranking of 1-10 MWe Solar Thermal Electric
     Power Systems, Vol.I, 1979 forthcoming. Golden, Colorado: Solar
     Energy Research Institute report SERI/TR-35-238.

D.S. Ward, Solar Heating and Cooling Systems Operational Results
        Conference, Summary, 1979. Golden, Colorado: Solar Energy
        Research Institute SERI/TP-49-209.
   Watt, On the Nature and Distribution of Solar Radiation, 1978.
        Washington, D.C.: Department of Energy Report HCP/T2552-01.
P. Weaver, Photobiological Production of Hydrogen, 1979. Golden,
        Colorado: Solar Energy Research Institute report SERI/TR-33-122.
H.L. Wegley, 1978. Siting Handbook for Small Wind Energy Conversion
        Systems, Richland, Washington: Pacific Northwest Laboratory
        Report PNL-2521.
W.T. Welford and R. Winston, The Optics of Nonimaging Concentrators,
        1978. New York, NY: Academic Press.
R.E. Wilson and P.B.S. Lissaman, 1974. Applied Aerodynamics of
        Wind Power Machines, Corvallis, Oregon: Oregon State University.
M. Yokell, Environmental Benefits and Costs of Solar Energy, 1979
        forthcoming. Golden, Colorado: Solar Energy Research Institute
        report SERI/TR-52-074.

Appendix I
A Partial Listing of
Solar Energy
Journals, Magazines, and Newsletters

SOLAR ENERGY TECHNOLOGY

Solar Energy (monthly). Pergamon Press, Headington Hill Hall,
    Oxford OX3 OBW, England. Technical journal with articles on
    wide variety of solar energy topics but with emphasis on solar
    heating and cooling. An official publication of the Inter-
    national Solar Energy Society.

Solar Age (monthly). Solar Vision, Inc., Church Hill, Harrisville,
    N.H. 03450, USA. Readable articles on good variety of solar
    energy topics. Emphasis is doing it yourself and solar heating.
    Official publication of the American Section of International
    Solar Energy Society.

Solar Engineering Magazine (monthly). Solar Engineering Publishers,
    Inc., 8435 Stemmons Freeway, Suite 880, Dallas, TX 75247, USA.
    Official publication of the Solar Energy Industries Association.
    Focuses on design, installation, and operation of solar heating
    and cooling systems.

Solar Energy Materials (quarterly). North-Holland Publishing Company,
    P.O. Box 211, 1000 AE Amsterdam, The Netherlands. New profes-
    sional journal on materials science research in solar energy
    conversion.

Journal of Energy (bimonthly). American Institute of Aeronautics
    and Astronautics, Inc., 1290 Avenue of the Americas, New York,
    NY 10019, USA. Technical journal with frequent articles on
    engineering research on solar energy and wind energy conversion.

Applied Solar Energy (bimonthly). Allerton Press, Inc., 150 Fifth
    Avenue, New York, NY 10011. Translation of the Russian language
    journal Gelioteknika. Technical articles with emphasis on
    solar thermal conversion.

WIND ENERGY

Wind Power Digest (quarterly). Michael Evans, 54468 CR31, Bristol,
    IN 46507, USA. Easy to read magazine primarily directed toward
    amateur wind enthusiasts.

Wind Engineering (quarterly). Multi-Science Publishing Co., Ltd.
    The Old Mill, Dorset Place, London E151DJ, England. Technical
    articles on wind resources and conversion systems.

BIOMASS AND PHOTOCHEMICAL CONVERSION

Resource Recovery and Conservation (quarterly). Elsevier Scientific
    Publishing Co., P.O. Box 330, 1000 AH Amsterdam, The Netherlands.
    Technical journal primarily on recovery and recycling of
    materials, but with good articles on energy from wastes.

Forest Ecology and Management (quarterly). Elsevier, P.O. Box 330,
    1000 AH Amsterdam, The Netherlands. International professional
    journal on management of forest ecosystems for energy and
    other purposes.

Journal of Photochemistry (monthly). Elsevier Sequoia S.A.,
    P.O. Box 851, 1001 Lausanne 1, Switzerland. Technical journal
    which has some articles on advances in photochemical conversion
    of solar energy.

International Journal of Hydrogen Energy (bimonthly). Pergamon
    Press, Headington Hill Hall, Oxford OX3 OBW, England. Official
    journal of the International Association for Hydrogen Energy.
    Includes articles on future energy systems based on hydrogen
    and technical articles on production of hydrogen from solar
    energy (and other sources).

OCEAN THERMAL ENERGY

OTEC Liaison (monthly). Popular Products, Inc., 1303 South Michigan
    Avenue, Chicago, IL 60605, USA. Newsletter primarily reporting
    on the US OTEC program.

RENEWABLE ENERGY POLICY

Solar Energy Intelligence Report (weekly). Business Publishers, Inc.,
    P.O. Box 1067, Silver Springs, MD 20910, USA. Emphasis is on
    politics of solar energy in the US, but also provides news
    reports on technical advances.

Solar Law Reporter (bimonthly). Solar Energy Research Institute,
    1536 Cole Blvd., Golden, CO 80401, USA. Legal journal on
    legislation and legal cases relevant to solar and wind energy.

CoEvolution (quarterly). Point, Box 428, Sausalito, CA 94965, USA.
    Informative and provocative articles and reviews on appropriate
    technology.

People and Energy (bimonthly). Institute for Ecological Policies,
    1413 K Street NW, Washington, DC, 20005, USA. News and essays
    representing radical views toward energy alternatives.

Technological Forecasting and Social Change (monthly). Elsevier
     North Holland, 52 Vanderbilt Avenue, New York, NY 10017, USA.
     Interdiscipilinary journal reporting social science research
     on alternative futures.

Futurist (monthly). World Future Society, P.O. Box 30369, Bethesda
     Branch, Washington, D.C. 20014, USA. Imaginative and sometimes
     provocative articles on world futures, with increasing
     emphasis on energy futures.

ENERGY FROM THE SEA WAVES

A. Blandino, A. Brighenti, and P. Vielmo

Tecnomare S.p.A.

Venice, Italy

INTRODUCTION

The debate on the utilization of energy resources, becoming wider and more dramatic since 1973, presents areas of uncertainty and controversial view-points all over the world.

However, it is the general opinion that a new consciousness is growing; the scarcity of oil is irreversible and destined to influence the world-wide development rate. This consciousness is acting practically by means of the new programs that many countries have already adopted or are going to adopt. These programs contain two basic statements: first, to control, to restrain and to rationalize the consumption of energy without over compressing the development of the production rate; second, to try to satisfy partially the energy demand by means of the so called (renewable and not) alternative sources, even if oil and gas seem destined, at least for a medium term, to play a fundamental role in the world-wide balance of energy.

Among these complimentary or integrative energy sources, it is necessary to distinguish between new applications that are to be considered as an improvement of sources already available, and the utilization of new energy sources that could constitute a net increase in the global energy balance. In both fields, anyhow, the level of the research and the possibility of practical applications are conditioned by different factors, such as the technological know-how, the cost-benefit balance, the environmental impact, the safety, the available energy content, etc. Keeping this in mind, it is very difficult to evaluate the potential development of every methodology, but it is evident that efforts must be increased in

121

every direction through an intelligent selection of various approaches and without looking for miracles.

In this general picture the research and development activities performed by TECNOMARE in the field of sea wave energy exploitation are more comprehensible. These activities concern the following three main aspects:

    (a)   Evaluation of the sea wave energy content.

    (b)   Interaction between the incident wave field and the absorbing device.

    (c)   Economic assessment of the wave power generator.

In the following, the main results relevant to these themes will be shown and discussed together with the computerized procedures by which they have been obtained.

WAVE POWER AVAILABLE

The exploitation of the sea wave energy content implies the knowledge of the wave characteristics as a function of the selected area, of the season, of the bathymetry, etc.

In other words, it is necessary to know a long term statistic of the above characteristics and the local bathymetry in order to have quantitative information on the wave power available.

In order to achieve this goal, TECNOMARE has developed the computer program ARDOC that furnishes the following results (see Figure no. 1):

    (a)   Energy and power content of the sea waves.

    (b)   Data (height, period and number of cycles) for the fatigue analysis of offshore installations.

    (c)   Forecast of the working time necessary to carry out marine operations.

The program ARDOC can also work as a data bank for bare meteo-oceanographical data.

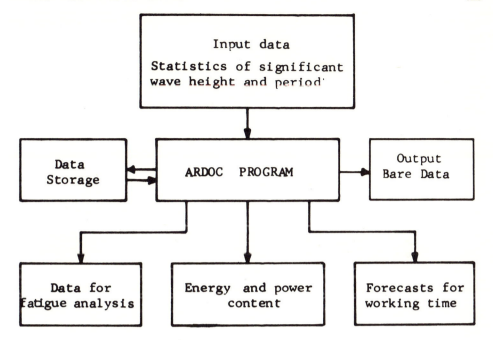

Fig. 1. Computer program "ARDOC"

The evaluation of the energy and power content of the sea waves is carried out through the following steps:

i) Determination of the real wave height and period statistics, using as input data the significant wave height and period statistics.

ii) Calculation, by means of the linear theory, of the mean power transferred in one period by a wave of height H and period T, through a vertical strip of unit width extending to the sea bottom.

The results relevant to the energy and power of the sea waves are:

i) Seasonal or annual tables (as Table 1) that, for each direction, give the percentages of the time during which the power content is within preassigned power bounds.

ii) Seasonal and directional tables (as Table 2) that give the power content in preassigned wave period and height intervals.

```
************************        TITLE      : GENERAL ENERGY TABLE                    023-A000-REL-G001
*   TECNOMARE S.P.A.   *
*   P R O G R A M      *        JOB NO.    : 623004              DATE : 10/05/77
*   081 - ARDOC-1      *
************************        ACTIVITY NO.: C704    EX'TD ? WP   PAG. 19
```

AVERAGE PERCENTAGE OF OCCURRENCE FOR POWER INTERVALS

AREA: 11.   NORTH SEA   LAT: 56 DEG. 32' N
DEPT: 230 FEET   LON: 3 DEG. 15' E
SEASON: ANNUAL

POWER INTERVALS (KW/M)	N	NE	E	SE	S	SW	W	NW	TOTALS
0 – 5	4.9	3.1	3.8	6.2	6.8	6.3	5.4	5.8	42.3
5 – 10	1.9	1.2	1.5	2.5	2.8	2.7	2.4	2.5	17.5
10 – 15	.7	.4	.5	.9	1.0	1.0	.9	.9	6.3
15 – 20	.6	.3	.4	.7	.8	.8	.4	.8	5.2
20 – 25	.3	.1	.3	.3	.4	.4	.4	.5	2.5
25 – 30	.4	.2	.3	.5	.5	.6	.6	.5	3.6
30 – 35	.3	.1	.2	.4	.4	.4	.4	.4	2.6
35 – 40	.2	.1	.1	.2	.2	.3	.3	.3	1.7
40 – 45	.2	.1	.2	.3	.3	.3	.3	.3	2.0
45 – 50	.1	.1	.1	.2	.2	.2	.2	.2	1.3
50 – 60	.2	.1	.1	.2	.2	.2	.2	.2	1.4
60 – 70	.2	.1	.2	.3	.2	.3	.3	.3	1.9
70 – 80	.2	.1	.1	.2	.2	.2	.2	.2	1.4
80 – 90	.1	0.0	.1	.1	.1	.1	.1	.1	.7
90 – 100	.1	0.0	.1	.2	.1	.2	.2	.1	1.1
100 – 200	.5	.2	.4	.7	.6	.7	.8	.8	4.7
>200	.4	.1	.4	.7	.4	.5	.6	.7	3.8
TOTALS	11.3	6.3	8.7	14.6	15.2	15.2	14.1	14.6	100.0

MEAN POWER (KW/M) : 36.
MAXIMUM POWER (KW/M) : 5983.
TOTAL ENERGY (KWH/M) : 312769.

NOTE: POWER AND ENERGY ARE REFERRED TO ONE METER CREST WIDTH

TABLE 1

```
*********************
*  -TECNOMARE S.P.A.-  *
*    P R O G R A M     *
*   081  -  ARDOC-1  -  *
*********************
```

TITLE : POWER TABLE

JOB NO. : 623004

ACTIVITY NO.: C704

023-A000-REL-G001

DATE : 10/05/77

EXITD : WP

PAG. 20

AVERAGE POWER VALUES (KW/M) FOR HEIGHT AND PERIOD INTERVALS

AREA: 11. NORTH SEA
DEPT: 230 FEET
SEASON: ANNUAL

LAT: 56 DEG. 32' N
LON: 3 DEG. 15' E
DIRECTION: ALL

HT INTERVALS (METERS) — PERIOD INTERVALS

	0-4	4-6	6-8	8-10	10-15	15-20	20-25	25-30	>30
	1.	6.	6.	4.	11.	17.	22.	29.	0.
	12.	19.	20.	37.	40.	71.	89.	93.	0.
	36.	56.	80.	103.	138.	193.	244.	0.	0.
	74.	119.	160.	205.	274.	377.	473.	0.	0.
	0.	206.	260.	342.	456.	617.	756.	0.	0.
	0.	304.	407.	515.	642.	914.	0.	0.	0.
	0.	430.	576.	724.	949.	1255.	0.	0.	0.
	0.	0.	781.	969.	1251.	1639.	0.	0.	0.
	0.	0.	1017.	1240.	1587.	2104.	0.	0.	0.
	0.	0.	1274.	1570.	1958.	0.	0.	0.	0.
	0.	0.	1647.	1920.	2344.	0.	0.	0.	0.
	0.	0.	1818.	2311.	2778.	0.	0.	0.	0.
	0.	0.	0.	2736.	3241.	0.	0.	0.	0.
	0.	0.	0.	3263.	3721.	0.	0.	0.	0.
	0.	0.	0.	3768.	4240.	0.	0.	0.	0.
	0.	0.	0.	0.	5253.	0.	0.	0.	0.
	0.	0.	0.	0.	0.	0.	0.	0.	0.

NOTE: POWER AND ENERGY ARE REFERRED TO 1 METER CREST WIDTH

TABLE 2

iii)   Seasonal and directional tables (as Table 3) that give
       the percentage of time (occurrence) in preassigned wave
       period and height intervals.

iv)   Seasonal and directional tables (as Table 4) that furnish
       the total energy associated with waves whose period and
       height are within preassigned intervals.

From Table 1 the importance of the directional characteristic
of the machine can be analyzed in the area considered. If the
device has some minimum power level sensibility, Table 1 gives also
quantitative information on the energy which cannot be recovered and
on the maximum yearly power which may occur on the marine system and
its supporting structure.

From Tables 3 and 4 it is possible to determine the theoretical
energy which can be extracted with respect both to the frequence
efficiency and to the wave height efficiency of the machine.

Following this kind of approach the usage factor can be determined. For example let's consider a marine device capable of taking
up and transforming the whole energy of the waves whose periods are
between T_1=6 sec and T_2=15 sec and whose heights are between H_1=2 m
and H_2=5 m.

In the selected area of the North Sea we could obtain 142.400
KWh/m every year (45.5 of the total energy) with the plant working
for 9.8% of the time.

INTERACTION SEA WAVES - POWER GENERATOR

The problem of extracting useful energy from sea waves involves
two main items (which are interactive): the device itself and the
sea.

Up to now the major effort of investigators and researchers were
devoted to the design of devices capable of somehow transforming the
motion of the sea into useful mechanical or hydraulic power without
much knowledge of the characteristics of the sea.

The true energy content of the sea and its distribution in time
has not been taken into special consideration compared with the
machine itself.

Regarding the machine itself, many possible devices have been
proposed, especially during this century. The present situation
can be summarized in this way: great approximation in approaching
the problem; a very large number of patients; many preliminary conceptual designs; and very few in-depth studies.

023-A000-REL-G001

PAG. 21

```
*******************
*  -TECNOMARE-S.P.A.-  *
*    P R O G R A M     *
*  081   AHDOC-1       *
*   ACTIVITY NO.: C704 *
*******************
```

TITLE : OCCURRENCE POWER TABLE

JOB NO. : 623004   DATE : 10/05/77   EX.TD : WP

AVERAGE PERCENTAGE OF OCCURRENCE FOR WAVE HEIGHT AND PERIOD INTERVALS

AREA: 11. NORTH SEA
DEPT: 230 FEET
SEASON: ANNUAL

LAT: 56 DEG. 32' N
LON: -3 DEG. 15' E
DIRECTION: ALL

HT INTERVALS (METERS)	0-4	4-6	6-8	8-10	10-15	15-20	20-25	25-30	>30	TOTALS
1	23.4	20.3	12.2	4.4	3.0	.3	0.0	0.0	0.0	64.8
2	.7	5.3	7.4	4.2	4.2	.5	0.0	0.0	0.0	23.3
3	0.0	.3	2.1	2.1	2.2	.3	0.0	0.0	0.0	7.5
4	0.0	.1	.4	.4	1.0	.1	0.0	0.0	0.0	2.6
5	0.0	0.0	.2	.3	.5	.1	0.0	0.0	0.0	1.1
6	0.0	0.0	.1	.1	.2	0.0	0.0	0.0	0.0	.4
7	0.0	0.0	0.0	0.0	.1	0.0	0.0	0.0	0.0	.2
8	0.0	0.0	0.0	0.0	.1	0.0	0.0	0.0	0.0	.1
9	0.0	0.0	0.0	0.0	0.0	0.0	0.0	0.0	0.0	0.0
10	0.0	0.0	0.0	0.0	0.0	0.0	0.0	0.0	0.0	0.0
11	0.0	0.0	0.0	0.0	0.0	0.0	0.0	0.0	0.0	0.0
12	0.0	0.0	0.0	0.0	0.0	0.0	0.0	0.0	0.0	0.0
13	0.0	0.0	0.0	0.0	0.0	0.0	0.0	0.0	0.0	0.0
14	0.0	0.0	0.0	0.0	0.0	0.0	0.0	0.0	0.0	0.0
15	0.0	0.0	0.0	0.0	0.0	0.0	0.0	0.0	0.0	0.0
20	0.0	0.0	0.0	0.0	0.0	0.0	0.0	0.0	0.0	0.0
>20	0.0	0.0	0.0	0.0	0.0	0.0	0.0	0.0	0.0	0.0
TOTALS	24.1	26.5	22.6	14.7	11.3	1.3	0.0	0.0	0.0	100.0

NOTE: - POWER AND ENERGY ARE REFERRED TO 1 METER CREST WIDTH
      - PERCENTAGES LESS THAN 0.05 ARE ASSUMED EQUAL TO ZERO

TABLE 3

```
*********************
*   -TECNOMARE S.P.A.-   *
*    P R O G R A M       *
*  081 -  A00C-1 -       *
*********************
```

023-A00C-REL-G001

TITLE : ENERGY TABLE

JOB NO. : 623004   DATE : 10/05/77

ACTIVITY NO.: C704   EXTD : 4B   PAG. 22

AVERAGE ENERGY VALUES (KWH/M) FOR WAVE HEIGHT AND PERIOD INTERVALS

AREA: 11.   NORTH-SEA
DEPT: 230. FEET
SEASON: ANNUAL

LAT: 56 DEG. 32' N
LON: 3 DEG. 15' E
DIRECTION: ALL

HT INTERVALS (METERS)

	0-4	4-6	6-8	8-10	PERIOD INTERVALS 10-15	15-20	20-25	25-30	>30	TOTALS
1	1704	6680	6080	3015	2940	417	34	1	0	20785
2	740	8939	14164	16410	18192	3100	247	8	0	66210
3	76	4089	14740	18653	26341	4444	287	0	0	69070
4	3	1503	8759	14317	25013	4416	149	0	0	54160
5	0	440	4607	9242	20049	2922	51	0	0	37961
6	0	91	2250	6462	14740	1537	0	0	0	25130
7	0	7	1022	4152	10141	635	0	0	0	15957
8	0	0	427	2663	6423	200	0	0	0	9913
9	0	0	170	1681	4064	27	0	0	0	5946
10	0	0	47	1013	2431	0	0	0	0	3491
11	0	0	12	644	1365	0	0	0	0	1971
12	0	0	3	312	746	0	0	0	0	1061
13	0	0	0	164	400	0	0	0	0	568
14	0	0	0	85	210	0	0	0	0	295
15	0	0	0	39	101	0	0	0	0	140
16	0	0	0	0	76	0	0	0	0	76
>20	0	0	0	0	0	0	0	0	0	0
TOTALS	2523	20749	66291	60706	133476	18107	773	9	0	312734

NOTE: -POWER AND ENERGY ARE REFERRED TO 1 METER CREST WIDTH

TABLE 4

Regarding the sea and its energy characteristics, the situation is the following: some difficulties, partly due to the mode-rate availability of useful statistical data and very few specific studies and theoretical means of analysis.

Nowadays, the highest level of development of the problem is reached when the available elements of the two main items are put together for an economic evaluation of the cost of the energy extracted (see Figure 2).

This preliminary approach is certainly very useful in early stage of development, but very approximate. In fact, the whole problem (waves and machine) must be regarded as a "system".

The assumption of neglecting the interaction between the wave activated device and the sea is unrealistic, even if necessary, in the earlier stages; the sea wave energy system cannot be divided into two (the wave activated machine with this characteristics, and the sea with its characteristics) as the two elements interact on each other. For this reason, it is necessary to seek a more organic approach as outlined in Figure 3 with a tentative block diagram.

The possible consequences of this approach are explained by the following considerations.

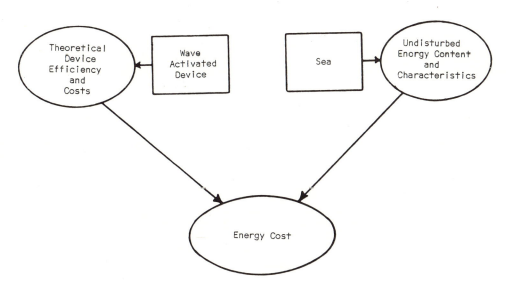

Fig. 2.   Preliminary approach to the problem of wave energy conversion

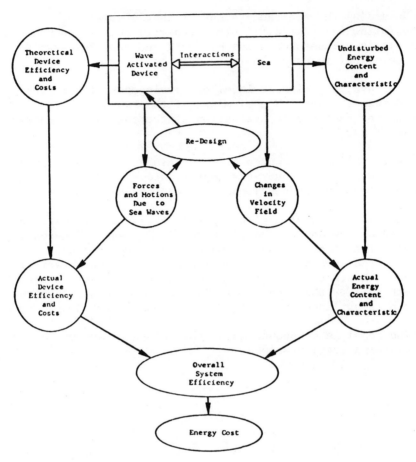

Fig. 3.   Improved approach to the problem of wave energy conversion

Generally speaking, it is not completely certain that the cheapest energy from waves can be obtained from the stormiest seas and deep waters where, nevertheless, the energy contained in the sea is greater than near the shore.

In fact, very rough seas imply additional high costs for the designing of the device and a strong structure which cannot be balanced by the increase of the power output. Also, for mooring, maintenance and power transportation, deep waters imply costs which may increase more quickly than the theoretically available wave power.

Another consideration linked to the improved approach to the problem of sea wave power, is that no special attention has yet been paid to the possibility of some kind of "pre-concentration"

of wave energy by means of the interaction effects between the sea
and the power absorber. This essential aspect has been developed
by us during the research, presently in course, sponsored by C.N.R.
(National Council of Researches).

THE POINT-ABSORBER CHOICE

The extraction of sea wave power needs a transferring process
from the sea to the equipment which must be able to absorb energy
from the velocity field originated by the waves, or from the pressure
field or from the free surface movements. The device itself modifies
the incoming wave field, perturbing considerably the wave amplitudes
in the surrounding zone (diffraction effect).

Figure 4 shows the equal value lines of the ratio of the dif-
fracted wave height over the incident one which supplies the variation of the disturbed crest elevation, at any point.

As the diffraction effects depend on the wave period the local
statistic of the waves, on which the total energy evaluation is
based, results modified in the sense that in some zones the energy
will be more concentrated, in others more rarefied. It is obvious
that the mean spatial assessment of the energy may be modified not
only by the useful location of the extracting device but also by an
appropriate geometric configuration, thus optimizing the mean overall efficiency of the system.

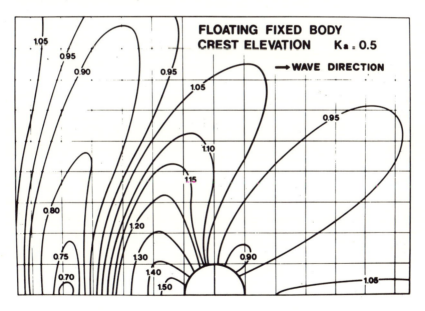

Fig. 4. Diffraction effects on the crest elevation

The system here analyzed (a vertical floating cylinder with an absorption device) is very simple but it is schematically representative of the whole family of wave energy extracting machines based on the large floating bodies' motions; they are usually called "point-absorber" devices.

The reasons for this choice are briefly summarized in the following considerations:

- As a result of the diffraction effects on the incoming wave field, caused by the extracting device, the intercepted energy is equivalent to that relevant to a crest length greater than the main dimension of the body. In other words it is possible to concentrate the energy by means of the diffreaction effects.

- By means of an appropriate assessment of a given number of floating bodies it is possible to obtain a structural weight less than a continuous structure, the KWh produced being the same in both cases.

- If the main dimensions of the extracting devices are contained, the reductive effect will be analogous for the wave loads acting on the bodies.

- A high degree of modularity produces great advantages both in terms of unit cost and in terms of maintenance.

- The modularity allows a major degree of freedom during the study and the realization of the possible disposition of the extracting devices in order to achieve pre-requested resonant conditions.

The system analyzed is composed of one or more vertical floating cylinders (rigidly connected) and is representative of the whole family of the machines based on the motions of large floating bodies.

PROCEDURE FOR THE EVALUATION OF THE SYSTEM EFFICIENCY

The general procedure for the calculation of the system efficiency is shown in the block diagram of Figure 5.

For each geometric configuration, a suitable mesh having been assumed, the most important part of the procedure consists in the use of our computer program DINDIF, which calculates the dynamic frequency response of a large body of arbitrary form on the basis of the three-dimensional diffraction theory.

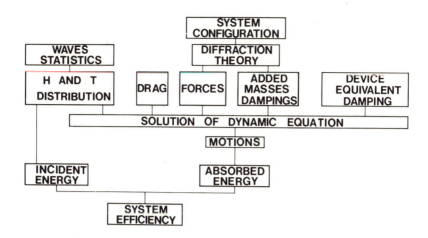

Fig. 5. Block-diagram of the procedure for system efficiency evaluation

The sink-source numerical technique is used to arrive at the solution. The results of the program include the following calculations:

- global forces and moment,
- added mass and damping matrix,
- linear motions in six degrees of freedom,
- dynamic pressure on the wet surface of the body,
- velocities, accelerations and pressures for points in the surrounding fluid.

In order to solve the linearized equation of the vertical body motion the following assumptions have been established:

- As it is necessary to consider the viscous drag resistance effects near to the resonance condition, a linearized heave damping coefficient has been added to the hydrodynamic one.

- The energy extracting device has been schematized as a linear absorber acting on the vertical motion.

In these hypotheses the total damping coefficient has been obtained as the sum of three terms:

a) the energy irradiated by the body,
b) the energy dissipated for viscous effects,
c) the useful energy absorbed by the device.

EFFICIENCY OF THE EXTRACTING DEVICE

The mean power transferred by a sinusoidal wave of height H and period T through a vertical strip of unit width is:

$$P = \frac{\rho g}{16\pi} H^2 \, T \, f$$

where

f = 1   in shallow water
f = 0.5 in deep water

The total energy (kinematic plus potential) transferred in a period T is:

$$E = PT$$

The total theoretical annual energy content in a geographical area is easily evaluated when the distribution of the number of waves (N_{jk}) into classes of H_j and T_k is obtained.

So the total annual incident energy is:

$$E_i = \frac{\rho g^2}{16} \sum_j \sum_k H_j^2 \, T_k^2 \, N_{jk} \, f_k$$

For each wave of characteristic H and T the energy absorbed by a machine with a linear absorber D_M is:

$$E_a = \frac{2\pi^2}{T} D_M \, X_3^2 \quad \text{(in one period)}$$

where X_3 = vertical motion amplitude.

If the system response is evaluated for the complete distribution of the waves (H_j, T_k), we can calculate the total energy extracted annually by:

$$E_a = 2\pi^2 D_M \sum_j \sum_k \frac{N_{jk} \, X_3^2 \, (H_j, T_k)}{T_k}$$

We assume as average annual efficiency of the system the ratio:

$$\eta = \frac{E_a}{E_i} = \frac{\text{Total annual extracted energy}}{\text{Total annual incident energy}}$$

where E_i refers to a wave front equal to the dimension of the body along the crest elevation.

This efficiency is the principal parameter to perform a comparison between different systems.

RESULTS OF THE PARAMETRIC ANALYSIS

A parametric period of dynamic analysis has been developed with reference to a vertical floating cylinder connected to a device with absorption D_M relevant to the heave motion. The parameters taken into consideration are:

- radius of the cylinder (from 5 m to 40 m),
- draft of the body (from 2 m to 16 m),
- equivalent damping (D_M) of the device,
- distance between the equal components (two or more) of a system.

For all the calculations a water depth of 100 m has been assumed.

Single Body System

The calculations based on the diffraction theory have allowed us to take into account the effective forces and hydrodynamic coefficients for each body. In fact, forces, added masses and damping coefficients are variable with the body geometry and with the period (see Figures 6, 7, 8).

For each geometric configuration a parametric analysis of the efficiency in function of the linear absorbing device D_M has been performed to find the value of D_M corresponding to the maximum average annual efficiency.

The importance of a correct choice of the linear absorber D_M is confirmed by the results plotted in Figure 9. The maximum global efficiency for the system examined is 22.2% and is obtained for an adimensionalized damping value $D_M/D_0 = 0.625$.

In the plot of the efficiency versus T (see Figure 10) it can be seen that for $D_M/D_0 = 0.2$ the efficiency reaches a peak value of 39% for T=9 s, which is approximately the resonance period.

But, as shown in Figure 9, for the value of D_M/D_O we have a low global efficiency (16%).

The curve of the global efficiency versus draft (see Figure 11), shows the existence of a maximum far draft = 3-4 m. Figure 12 shows also the dependence of the global efficiency on the radius.

For values of the radius greater than 20 m, the increase in the efficiency is very weak and the best zone seems to be between 20 and 30 m. Also in these curves for each value of the radius, the absorber D_M value has been optimized.

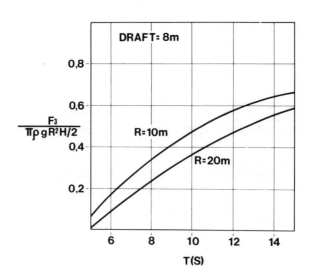

Fig. 6.   Vertical force as function of wave period

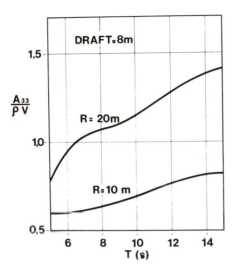

Fig. 7.   Added mass as function of wave period

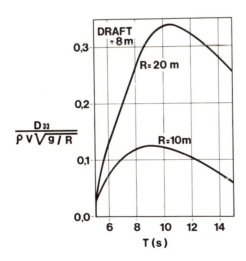

Fig. 8.   Damping coefficient as function of wave period

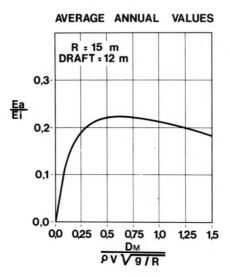

Fig. 9.   System global efficiency as function of
          damping coefficient

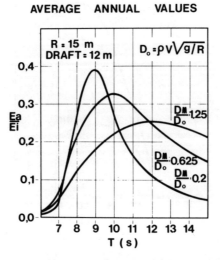

Fig. 10. System efficiency as function of
          wave period

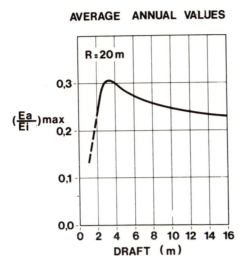

Fig. 11. Max global efficiency as function
          of draft

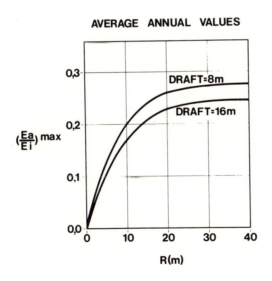

Fig. 12. Max global efficiency as function
          of radius

Multibody System

The results obtained for a system composed of two, three and four rigidly connected cylindrical bodies, are shown in Figures 13 and 14.

The overall efficiency is plotted as function of the mutual distance D, the maximum efficiency relevant to a single body system is also shown in order to make the comparison easier.

Figure 13 shows how the mutual interaction can increase the average global efficiency of each single component of a multi-body system; furthermore, the single body geometry being the same, an optimum number of bodies exists; in our case the maximum efficiency is obtained when the distance between the bodies is 15-25 m and for a three body configuration.

In Figure 14 the efficiency of such a system is shown as function of the ratio of the real mass M over that corresponding to the displaced volume (Mo). When the ratio is less than 1, the global efficiency decreases as the system is tuned mainly to the high frequencies where the energy content is scarce. On the contrary, when the ratio is greater than 1 we obtain the opposite effect as in low frequency bands the energy content is high.

The above results have confirmed that one of the most important problems concerning the extraction of energy from sea waves is a proper accurate parametric analysis to obtain the maximum efficiency by means of the diffraction theory. The main parameters which must be taken into account in this optimizing process are:

- typical dimensions,
- equivalent damping of the absorbing device,
- system composed of one or more components,
- number of the components,
- mechanical characteristics of the components.

The results shown in Figure 14 have demonstrated that an interesting possibility of optimizing consists in a device able to tune itself on the frequency of the incident waves. In this case it is reasonable that the global efficiency can increase to vary high values.

It is obvious that such a system will be analyzed from a technical point of view since, in order to realize a self-tuning system operating in a marine environment, it is necessary to solve a long hard series of problems concerning the reliability, the durability and the cost of the machine.

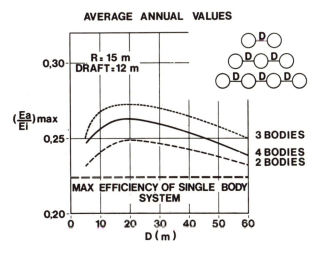

Fig. 13.   Max efficiency of multi-body system
           as function of mutual distance

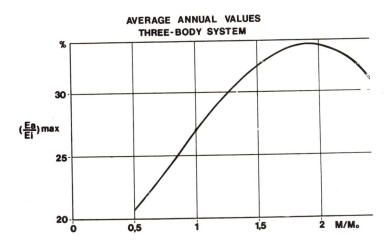

Fig. 14.   Max global efficiency as function of the ratio of the
           real mass M over that corresponding to the displaced
           volume (M_0)

ECONOMIC ASPECTS OF WAVE ENERGY PRODUCTION

The cost per KWh of the power generated by various wave energy devices under given nominal conditions will be presented in the following.

Besides the cost level as such, there are some further aspects which must be considered due to the specific features of wave energy power.

Wave energy is a "low quality" energy due to its varying and uncontrolled occurrence, and often, also, its remote location. Some of the advantages and disadvantages of wave energy generated power are the following:

- The uncontrolled occurrence makes supplementary power supplies necessary.

- Adjusting the irregular output to an existing distribution grid is relatively expensive.

- The variations are too big for direct power supply to most types of offshore processing plants; hydrogen production via hydrolysis is probably the only one able to accept these variations.

- The production increased during the winter season when the demand is also higher.

Since wave energy is always of very low density, the size of any device tends to get very big and consequently, expensive in relation to the power that can be expected.

The cost of the structure will always be much higher than the cost of the power generating equipment. A device with an acceptable cost situation must thus have as little as possible static structure and essentially work on dynamic principles.

This aspect may be visualized by a simple calculation. If the energy available is 50 KW per meter crest length, and assuming 15% excess loss and 35% total efficiency, about 15 KW/m can be generated.

If the approach is taken that the device may cost as much as a typical hydro-electric plant, about 700 $ US per mean KW, some estimates may be made of the acceptable cost of the structure. If the cost of the power generation, transmission and mooring of the device is assumed to be one third of the total cost then about 450 US $ per KW may be spent on the structure; the structure cost per meter crest length may be 15 x 470 = 7000 US$.

If the structure is built in steel at a cost of about 1800 US$
per ton, a one-meter section of device may contain as a maximum
= $\frac{7000}{1800}$ = 4 tons of steel.

This obviously rules out all devices containing large pontoons
or other floating devices the main dimension of which is a full wave
length, if they are to remain stationary. Only devices based on the
principle of moving the waves have some possibility of satisfying
this weight target criterium.

Such considerations show that it is very difficult to find
configurations having a sufficiently small cross section but still
able to absorb the power. The studies presented here clearly show
how important the weight and cost of structure are for the overall
result.

For certain types of devices there may be a scale factor limiting the economic size of the device. Device buoys which, when scaled
up under given wave conditions grow three-dimensionally will produce
a cost increase almost in proportion to the third power of the size
whereas the wave absorbing capacity increases only in proportion to
the size. Moderately sized units in larger members may therefore be
more feasible than very large ones.

ECONOMIC EVALUATION OF A WAVE POWER PLANT

A procedure of calculation has been utilized to evaluate the
conditions of profitability of some sea wave energy plants, or better,
to determine the minimum conditions for which the investment is
capable of yielding an interest at least equal to the minimum acceptable net rate of return. The following hypotheses are valid as input:

- Average annual available power   36 KW/m (North Sea)
- Overall efficiency   35 %
- Operating life of the plant   10 years
- Time required by design, build-
  ing, installation   2 years
- Maintenance and operative costs   10 % of the investment
                                    per year
- Over-head costs   1 % of the investment
                                    per year

The total estimated investment, per meter of wave crest, will
range within 20,000 ÷ 60,000 $/m of wave crest.

The results obtained show that the above-mentioned profitability
conditions are satisfied only if the sales price of the energy at the
production start-up is superior to 10.5 ¢/KWh, assuming a nominal
investment of 40,000 $/m. Table 5 shows a forecast of the cost of

solar, nuclear, and coal-burning energy (Ref. 3). The estimated costs of all these sources of energy exceed 10 ¢/KWh, should the production start in the year 2000. This allows us to assume that the sales price of KWh will reach that level by the end of this century, also making a profitable exploitation of the sea wave energy.

CONCLUSIONS

Although the principle of the wave energy conversion concept was proposed more than a century ago, many scientific, technical and economic aspects still have to be clarified before any industrial development can be seriously envisaged.

Today many ideas and projects are emerging and have been proposed in various parts of the world. However, it is necessary to emphasize that the conclusions so far reached have to be considered as provisional since this brief analysis still needs to be assessed in the proper economic and political perspectives.

From a structural/constructional point of view wave energy systems in general offer the possibility of using various steel and concrete products selected on the basis of an optimal assessment of labour, materials and energy costs. Although quite a large number of potential sites for wave energy conversion systems around Europe are mentioned in the lieterature, only a few sites exhibit a sufficient energy content and are located in areas suitable for possible industrial operations.

Furthermore, it is possible to identify additional potential sites that require further wave data collection and processing before any definite conclusions can be reached.

In general we have the feeling that at the present stage of development, industry and specialized societies could use ther experience and competence in order to assist and to cooperate with universities during the development and fulfillment of their programmes. Such an approach would provide an opportunity of each other direct access to first-hand, practical information and of becoming involved in the already existing research and development activities.

Table 5 - Comparison of Energy Cost

	Load Factor	ENERGY COST, ¢/kWh-e				
		Capital	Other[2]	O & M	Fuel[3]	Total
SOLAR						
1975 START-UP	0.54	4.29	3.60	0.92	0.30	9.1
2000 START-UP	0.54	5.67	4.76	1.15	0.57	12.1
NUCLEAR						
1975 START-UP	0.55	1.08	1.17	0.23	1.17	3.7
2000 START-UP	0.70[4]	4.19	4.55	0.23	1.47	10.4
COAL						
1975 START-UP	0.62	0.93	0.96	0.39	2.85	5.1
2000 START-UP	0.74[4]	3.70	3.84	0.48	5.74	13.8

1. 1975 DOLLARS

2. OTHER INCLUDES INSURANCE, PROFIT, TAXES ETC.

3. FOR SOLAR PLANTS, FUEL INCLUDES BACKUP CAPACITY AS WELL AS FOSSIL FUEL REQUIRED FOR MARGIN

4. IMPROVED LOAD FACTOR ASSUMED POSSIBLE BY AD 2000

NOMENCLATURE

A_{33}	Heave added mass for single body system
C_D	Drag coefficient
d	Draft
D	Mutual distance of two bodies in a multi-body system
D_M	Device damping system
D_{33}	Heave damping coefficient for a single body system
D_0	$\rho V \sqrt{g/R}$
E_a	Absorbed energy
E_i	Incident energy
f, f_k	Incident energy factor
g	Gravity acceleration
H, H_j	Wave height
K_a	Dimensionless wave number
M	Real body mass
M_0	Body mass, corresponding to the displaced volume
n_j	Surface generalized normal
N_{jk}	Number of waves into classes of height H_j and period T_k
R	Radius of the body
s_j	Motion amplitude (j = 1,6)
T, T_k	Wave period
V	Displaced volume
η	Average annual efficiency
ρ	Mass density of water

REFERENCES

1. Berta, M., Blandino A., Paruzzolo A., 1979, "An integrated
    procedure to compute wave loads on hybrid gravity plat-
    forms", International Conference on Environmental Forces
    on Engineering Structures, Imperial College, London, July.

2. Berta M., Blandino A., Marcon D., Paruzzolo A., 1979, "Sea-
    structure interaction for tripod type steel gravity plat-
    forms", Brasil Offshore 1979, International Symposium on
    Offshore Structure, Rio de Janeiro, October.

3. Caputo R.S., Truscello V.C. (J.P.L. Pasadena, Cal.), 1976,
    "Solar thermal power plants: their performance character-
    istics and total social costs", 11th Intersociety Energy
    Conversion Engineering Conference, State Line, Nevada,
    September.

4. Faltinsen O.M., Michelsen F., 1974, "Motions of large struc-
    tures in waves at zero Froude number", Proceedings of
    International Symposium on the Dynamics of Marine Vehicles
    and Structures in Waves, University College, London.

5. Garrison C.J., Berklite R.B., 1972, "Hydrodynamic loads in-
    duced by earthquakes", OTC 1554.

6. Garrison C.J., Chow P.Y., 1972, "Waves forces on submerged
    bodies", Journal of the Waterways, Harbors and Coastal
    Engineering Division - ASCE, August.

7. Kinsman B., 1965, "Wind Waves", Prentice-Hall.

8. Leishman J.M., Scobie G., UK N.E.L., 1975, "The development of
    wave power - A techno-economic study".

9. Loken A.E., Olsen O.A., 1976, "Diffraction theory and statisti-
    cal methods to predict wave induced motions and loads for
    large structures", OTC 2502.

10. Perry J., Davies S., 1979, "UMIST scrutinises economic apprai-
    sal of offshore construction", Offshore Engineer, April.

11. Sebastiani G., Berta M., Blandino A., 1978, "Energy from sea
    waves: System optimization by diffraction theory", OCEAN 78,
    Washington, September.

THE ROLE OF ELECTRIC SYSTEM IN THE DIVERSIFICATION OF ENERGY

SOURCES

Luigi Paris

Manager R&D Department
ENEL Italian Electricity Agency
Via G. B. Martini, 3
00100 - ROMA - Italy

1. GENERAL REMARKS

As you know I work for a big electric utility; therefore I
apologize for viewing energy problems as a specialist in electricity, which plays today in the energy system an important role of
energy carrier from the source to its final uses. In opposition to
those who see in electricity and the electric system a barrier to
the introduction of new possible power generation and consumption
models, capable of coping with the energy crisis, I see in them a
fundamental tool to solve this crisis. Electricity can contribute
to energy source diversification and thus exploit energy sources
alternative to oil and gas. It is useless to remind the reasons
why nuclear energy can be used today only by converting it into
electricity. This is perhaps the major fault of electricity: to
make the use of nuclear energy safe, economic and possible. But,
even if our society wishes to throw away the great possibility that
nuclear energy offers to solve the energy crisis, a possibility
that he secured with his ability and his work, I do not see why he
should throw away electricity too. On the other hand, coal, the
other large alternative source, has some trouble in replacing oil
and gas in the majority of current applications due to its difficult handling and combustion; therefore it must be transformed into
fluid or more economically must be converted into electric energy
in large power plants. Solar energy, the third alternative source,
seems to be the only large source that can be used without using
the electric carrier, because it is everywhere available, and can
be converted into a suitable form wherever desired. The other un-
transferable energy sources (like hydroelectric, high-enthalpy

geothermal, wind energy and other renewable sources) hardly ever
find an immediate utilization on the spot and can thus be used only
through electricity.

However, what makes the electric carrier particularly interesting with regard to source diversification, is the fact that all
energy sources can be made accessible to the electric user.

This characteristic is very significant. In fact, if a variation in the policy of source utilization involves a modification in
the small and numerous units using such sources, this modification
would be difficult to carry out and only on a long-term basis. The
reason for this is that substantial alterations in extensive service
structures and, above all, a change in the habits of millions of
people would be necessary.

The diffusion of electric users thus allows maximum flexibility
in the policies of source differentiation, always provided, of
course, that the electric user makes a proper use of this high-grade
energy.

Let's go back for one moment to solar energy and to its gift of
ubiquity, which makes it possible to replace oil and gas in small
thermal utilizations like home heating.

Evidently, in this case, the use of an electric carrier would
be neither necessary nor, especially, advantageous. In fact, the
direct conversion of radiant energy into heat takes place with a
good efficiency in relatively simple and economic apparatus.

Similarly, one can think of converting solar energy into electric power to supply the conventional electric power users in situ,
possibly be combining generation of electric power and heat. The
idea which naturally follows is an electric system which resorts to
a myriad of small generators, rather than to large generators which
feed a myriad of small users.

Such philosophy is feasible even when the primary source of
small thermal users is not the sun but conventional fuels or, better,
local fuels (like biogas, waste, etc.); in this case, small combined
production of power and heat may contribute to equip our system
with small distributed generations.

The interest in this type of distributed generation system is
based in part on pretended large savings that could be obtained in
the design of the transmission and distribution network and in its
management or, even more hypothetically, in the elimination of such
network.

On the other hand, an economic operation of such a distributed system requires an electric system performing some unreplaceable auxiliary functions.

These functions are:

Reserve

In order for the local electric energy units to guarantee the levels of continuous supply required today, they should have reserves which would excessively increase their costs. By making the reserves available anywhere, the electric system can enable a high utilization of such reserve and thus lower costs enormously.

Integration of Loads

Individual power requirements generally have a very irregular pattern over time and different from one another. Their integration thus allows a more regular demand thereby increasing utilization of the power-producing units and sizeably lowering their cost.

Integration of Generations

The different generation systems involved in a diversified system of energy sources can be to a greater or lesser extent restricted in their production by external conditions (astonomic and meteorologic factors of various nature for hydroelectric, wind and solar energy), heat production needs in heat productions, combined productions, etc.; or by economic conditions of better exploitation.

It is evident that through the integration of sources, it is possible to obtain enormous savings both as regards plant design and operation.

Storage

Significant efforts in energy storage are essential in a system that is based on energy sources dependent on astronomic and meteorologic events. In effect, the electric system enables one to obtain such storage in the most favorable conditions, freeing the storage systems of location constraints.

Therefore, distributed generation is a sound idea but cannot be paid by eliminating the electric system; it follows that the size and the number of generating plants is a pure matter of conversion cost.

After these general remarks on the present validity of electricity and of the electric system in the framework of energy crisis, my lectures will be divided into three main topics.

First, I will discuss the consumption of the basic resource for the production of electric power. Then, I will deal with competitiveness comparison between energetic technologies related to electricity. Finally, I will deal with storage of mechanical and electric energy as a basic problem in energy conservation.

2. BASIC RESOURCES FOR ELECTRICITY

Introduction

The events in the last decade, referred to as environment and energy crises, produced several changes in engineering thinking habits.

One of the most evident aspects of these changes is the lack of confidence in economic optimization as a basic decision tool, capable of ensuring the best management of resources. An attempt was made to find a replacement of this decision tool in the large, universal, macro-economic models. But, because of their complexity, they risk to remain centralized tools in the hand of a cultural elite while they cannot be conveniently used by the great number of technologists, engineers and researchers for checking the solutions under study.

Today, therefore, in the energy debate somebody says that a given way of producing or consuming energy is "appropriate from the energy viewpoint" or "appropriate from the environmental viewpoint"; we are less interested in the fact that such a way of producing or consuming energy is "appropriate" in economic terms.

This is misleading, makes choices highly debatable, and generates difficulties in understanding the mutual positions; moreover, it determines a cleavage between innovative thinking and reality since our system rewards only economic choices.

Minimum Resources Utilization Criterion

To revalue economic optimization as a resource tool, we should consider that any product and any service supplied to man consumes or utilizes other products or services; but at the beginning of the chain we always find the following three basic resources (fig. 1):

raw materials,
labor,
environment and land.

Fig. 1.   The three basic resources.

As it is possible to secure the same product or service by using different combinations of these resources and since they are available in limited amounts, it is absolutely necessary to choose the approach which minimizes their use on the whole.

To get a minimum overall use of so different quantities, it is necessary to measure them with the same unit; this unit has been money for centuries and the minimum operation is called economic optimization (fig. 2).

The universal value of this logic tool is unquestionable. On the other hand, we realize that long-term choices made exclusively in current economic terms may involve big strategic errors. Indeed, the current prices of some basic resources (such as land or energy resources) may be inadequate for their value and not take into account their progressive exhaustion.

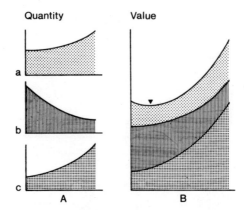

Fig. 2.   Each product or service uses a different combination of the
three basic resources: raw materials, labor, environment
and land (a); to get a minimum overall use of these resources, they must be measured with the same unit,
money (b).

While it is absolutely undesirable to neglect the economic optimization criteria, we should, in economic evaluations, give the
money value a more general meaning, which does not necessarily
correspond to the current price.

For instance, if we are afraid that the market value of a given
resource (e.g. oil) is unsuitable in relation to the consequences
of its consumption on the balance of payments, it is unwise to drop
any economic assessment and to limit ourselves to energy assessments. It is preferable, once again, to make use of economic
assessment assigning such resource with a fictitious strategic
value accounting for the dreaded consequences.

The Gradual Penalization Approach

These criteria must be applied also to the resouces which are
generally considered as not quantifiable and economically apprais-
able.

A topical question concerns today the measure and the economic
evaluation of the environmental alterations brought about by the
energy conversion, transmission and utilization system.

To this aim we have to set aside present criterion of strict
limits to environmental alterations and replace it with a gradual

strategic penalization corresponding to the effect brought about by each alteration.

The usual present method consists in fixing strict limits beyond which the environment is regarded as altered and below which the alteration is regarded as non-existent.

Though such simple and accessible criterion proves handy in solving possible disputes between utilities and control authorities, it is rough and unsuitable to optimize the use of limited resources available.

In effect, in some cases the possibility of reducing an alteration below the limits is disregarded, even when feasible with a modest use of other resources; in other cases a remarkable use of resources is likely to be necessary to reduce an interference even slightly, but below the conventional limit.

It is worthwhile to spend a few words to show how gradual penalization could work in the electric energy system.

Considering that the nuisance due to the polluting effluents of a power plant (or that due to the impact of a high voltage overhead line) reduces land useability for a given purpose and the extent of such reduction depends on the extent of the nuisance, we propose to associate each level of nuisance with a penalty factor which reduces the worth of the land (fig. 3).

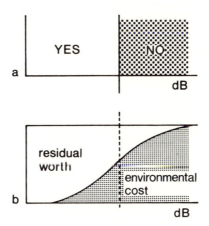

Fig. 3.   Comparison between the present "strict limit" method and the "environmental cost" method: a)   above the limit the environment is regarded as altered and below the limit the alteration is regarded as non-existent, b)   a gradual penalization corresponds to the environmental cost.

Since the same environmental effect produced by the line has different consequences depending on land utilization, such utilization conditions the assessment of these penalty factors. It will thus be necessary to work out an appropriate classification of the land into categories.

Of course, the assessment of a gradual penalization to be assigned to a given measurable nuisance is at present highly arbitrary, but no more than the choice of a limit.

Having determined the penalty factors for the various nuisances and for each land category, they will be suitably added up to obtain an overall penalization factor that enables one to evaluate the loss of the land value brought about by the power plant or by the overhead line. It will thus be possible to determine whether a measure taken during the design stage to attenuate the disturbing effect, and which causes a certain increase in the plant cost, is offest by a lower land devaluation.

Further research and experimental work have to be done with a special emphasis on the evaluation of land (or water, air quality, etc.) worth reduction due to each environmental alteration, in order to achieve an acceptable assessment of this criterion.

Universal macroeconomic models, mentioned before, may help to set up this kind of strategic pricing.

Analysis of Basic Resources Content of Electricity

We are working in order to verify if the application of such criterion is feasible in the energy sector. For the time being, due to the lack of many data and to the difficulties in assessing environment penalizations, it is impossible to use them in making choices; anyway, along these lines, it is possible to shed light on some basic points which govern energy policies.

For instance, when we speak about energy conservation, do we realize which basic resources we want to conserve energy among raw materials, environment and labor? In fact, all actions proposed to conserve energy consume a mix of basic resources; to be able to judge among different actions we have to know the basic resource content of the various forms of energy.

In this first lecture, I will limit myself to illustrate an analysis of the basic resource content of the various forms of energy which range from crude oil to electric energy supplied to domestic users and generated in thermal power plants.

This analysis starts from a first evaluation of basic resources, based on 1977 prices, for raw materials and labor and is regardless of a direct incidence of environmental costs (even if this evaluation reflects today's environmental difficulties through some prices for raw materials).

Fig. 4 indicates the basic resource content of energy in all

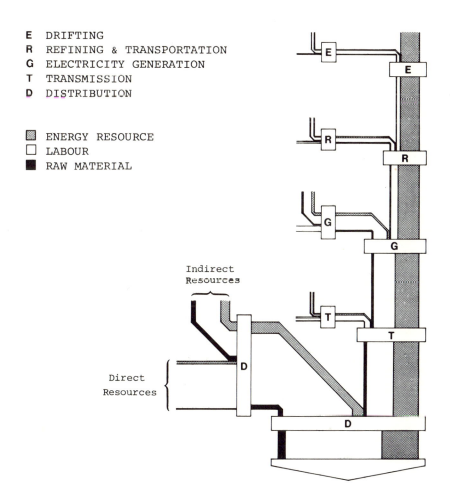

E   DRIFTING
R   REFINING & TRANSPORTATION
G   ELECTRICITY GENERATION
T   TRANSMISSION
D   DISTRIBUTION

ENERGY RESOURCE
LABOUR
RAW MATERIAL

Fig. 4.   Basic resources content of electricity produced from crude oil.   At each stage, the upper share is referred to indirect resources, while the lower share is referred to direct resources.   No environmental penalization considered.

the forms considered, i.e.:

    1.   crude oil
    2.   fuel oil
    3.   electric energy at the power plant
    4.   electric energy supplied to large industrial consumers
    5.   electric energy supplied to domestic users.

These energy forms are obtained through the following processes:

    -   extraction
    -   refining
    -   electricity generation
    -   electricity transmission
    -   electricity distribution

In this first picture, the basic resources are quoted on the base of actual market cost. Conventional environmental costs are therefore disregarded.

Since, among raw materials, the energetic raw material is of special interest, in our analysis it was particularly evidenced.

We extended the analysis of basic resources also to the numerous raw materials (steel, aluminum, cement, etc.) involved at the various stages; therefore, for instance, in fig. 4 the share of labor and energy resource used in the manufacture, transport and processing of steel used in electric transmission appears aggregated with labor and energy resources involved in the whole transmission stage, while the pure raw material share is represented by the market value of raw materials (coal, iron) used in steel manufacturing.

It is important to note, at this stage, how in this diagram we assigned a value to the energetic basic resource.

We deducted from the final crude oil market price, including royalties of the producing country and taxes of the importing country, the costs for extractions and transport to refinery, which were evidenced as consumption of labor, energy resource and non-energetic raw materials.

Since the crude oil extracted has a market value related only to quality and not to difficulty and cost of extraction, with an assessment of this type the value of the energetic raw material is lower where the extracting costs are higher.

This produces some difficulties in the parametric analysis as a function of resource value; for this analysis, we should necessarily refer to the value of extracted resources but with a given origin.

In fig. 4 we refer to an oil with a relatively high extraction cost (North Sea oil).

When dealing with economic evaluation, capital is currently considered as a basic resource; in our analysis, we take it into account by considering its influence in each process in terms of the above defined four basic resources.

At each stage of the energy conversion and transmission chain the content of resources increases; such increase is determined by an amount of resources directly consumed in the above process (lower share of the input resources at each stage; see fig. 4) while another part is consumed only indirectly (upper share). The indirected resources are those necessary for building the plant and the tools required for carrying out the processes (extracting or drifting equipment, transport means, refineries, electric power plants, lines, etc.).

Of course, it is evident that at each stage the energy process must include only a share of the resources consumed to build the relevant plant; this share financially corresponds to the depreciation charge of the plant during its lifetime.

But it is also appropriate to take into account the fact that we must penalize somehow any earlier consumption of resources in relation to the actual period of use of the energy produced; this financially corresponds to the application of an interest rate on capital.

It could be discussed about the application of this penalization and if this must be applied only to labor or even to raw materials; anyhow we applied it to the basic resources, in a manner variable but equal for all of them; in particular, in the diagram of fig. 4, it is applied at 5% per year.

The content of energetic resource in the electric energy supplied to domestic users is modest; to conserve energy at this level means to save labor rather than energy resource. Obviously, the situation of energy at fuel-oil level is quite opposite; here, indeed, any saving regards mainly the energetic raw material.

For electric power at generation or large industrial consumer level, the situation is intermediate with an almost equal share between oil and labor.

Starting from this analysis, it is possible to see what happens if we assign the energy resource a strategic value higher than the market value, so as to decrease its consumption in view of its exhaustion. Fig. 5 represents a borderline case where the price of

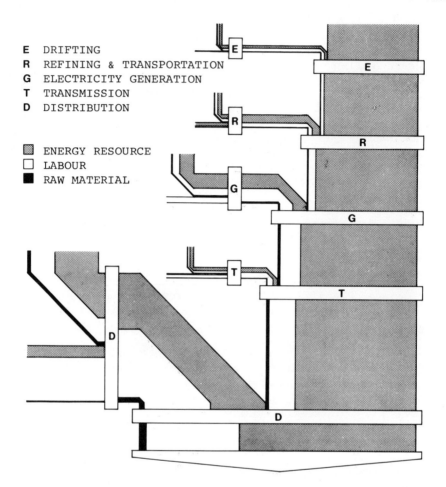

Fig. 5. Basic resources content of electricity when energetic raw
material price is increased 5 times more than labor and
other materials.

energy resources is increased 5 times more than labor and other raw
materials. In this situation, energetic raw materials are prevalent
even in electricity at domestic users.

Such an increase in energy resource value automatically entails
the use of other energetic raw materials more available than crude
oil, but which imply a higher use of other resources and namely

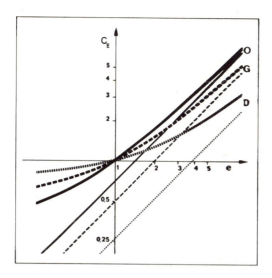

Fig. 6.   Energy cost C_E at Oil, Generation and Distribution stages
plotted versus e, which is the ratio of energy resource
unit value to the unit value of other resources.

labor for extraction, handling, and conversion into electric energy,
such as oil shale, coal or nuclear energy.

In can be concluded that as we increment the energetic raw
material value, electric energy is bound to consume more and more
labor resource.

This conclusion is supported by Fig. 6 where energy costs at
the various stages are plotted against the ratio of the energy resource unit value to the unit value of other resources (labor and
other raw materials).

The consequences of these variations are less evident when getting further from the primary source in the conversion and transport
energy chain.

Electric energy at distribution level is indeed much less sensitive to the cost of energetic raw material than the fuel oil.

It is also interesting to see what happens when varying the
penalization due to earlier resource consumption (interest rate).

Fig. 7 shows: a)   the variation with interest rate of the
percentage of total cost due to energy resource at the various levels;
b)   refers to the case of an increase for the energy resource five
times higher than that of other resources.

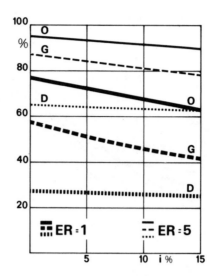

Fig. 7.   Percentage of total energy cost due to energy resource at
          Oil, Generation and Distribution stages as function of
          interest rate.   Thicker lines refer to the case of energetic
          raw material price increased 5 times.

The Assessment for a Penalization for Environmental Deterioration

        All the above is independent of the environment resource; no
doubt that energy consumption is seen with concern in view not only
of primary energy sources exhaustion, but also of land occupation and
environment deterioration, which today represent the major constraint
to the development of electric power systems.   And here the problem
becomes much more difficult; in effect, the environment and land
resource is hardly quantifiable in absolute terms.   Only for complete
land occupation can a useful market evaluation be resorted to.

        When the environment surrounding the plant is in some way altered
and quality and useability of the land resource is reduced, the
assessment of an economic penalization becomes much harder; in some
cases this can be done by quantifying the damage caused; in other
cases evaluating the reduction in the land market value, but often
it is necessary to introduce a conventional penalization.

        Furthermore, when the environment disturbance has a negligible
local effect but contributes, even to a minor extent, to affect
the present general balance of our biosphere (like the increase of
CO_2 density in the atmosphere due to combustion processes) only

conventional penalization can be used.

Similar approach applies when account must be taken of very low risks of very dangerous accidents (like the evaluations of the consequences of a serious accident in the operation of a nuclear power plant).

However, even when the choice of the environment penalization values is highly arbitrary, the penalization criterion is still valid; if we conventionally establish the penalization value we will still make choices, perhaps not the best but certainly consistent with one another, and over time, we should be able to adjust the penalization value to satisfactory levels.

It is always possible to assume the economic value of an environmental alteration as a parameter in a sensitivity analysis, so as to examine the consequences of its different evaluation in the policy choices to be undertaken.

Evaluation of Environmental Cost

In order to assess the consequences of a given alteration, we must first of all measure it; this is often very difficult, as in the case of landscape alteration, which represents the most significant nuisance brought about by overhead transmission lines.

Table 1. Measure and Penalization of Environmental Nuisance

NUISANCE	DIFFICULTY DEGREE	
	MEASURE	ECONOMIC PENALIZATION
SPACE OCCUPATION: AT GROUND	*	*
SPACE OCCUPATION: OVERHEAD	*	* *
IMPACT ON LANDSCAPE	* * *	* *
FALLOUT OF HARMFUL EFFLUENTS	* *	* *
PHYSICAL AIR & WATER ALTERNATIONS	* *	* * *
NOISES RADIOINTERFER OTHERS	*	* *
GLOBAL ALTERNATIONS	*	* * *

Table I reports the various nuisances to the environment caused by electric generation, transmission and distribution systems as well the degree of difficulty in both measuring and economically penalizing each nuisance.

Space Occupation

Space occupation requirements are easily determined considering the land which, because of system operation or for safety reasons, cannot be utilized for other purpose.

Space occupation cost can be easily determined from expenses for purchase of lands and right-of-way acquisition.

Impact on Landscape

This nuisance is today considered the most difficult to quantify; even if many proposals have been put forward, we feel that most seem to lack real self confidence in their effectiveness and therefore need further discussion and improvement in order to be commonly accepted.

A measurement of such alteration which we propose is based on the geometric evaluation of the occupation of the visual field by the disturbing element, corrected with some coefficients taking into account the extraneousness of the element in the landscape and other psychological effects that can be evaluated through opinion surveys.

Then we quantified the nuisance to the landscape with criteria very arbitrary in absolute terms but very accurate in relative terms for the different components of the electric system.

To give an idea of this first-hand assessment of nuisance to landscape, you may consider that the cost of this disturbance, attributed to the power plant of figure 8, was evaluated equal to about 10 times the purchase cost of land for its construction.

Polluting Effluents

We gave particular consideration to sulphur oxides emission; for fuel oil thermal power plants we derived the measure and the costs of such emission from the evaluation of the damages due to combustion processes over all the national territory and the definition of the share to be attributed to thermal power plants.

Fig. 8.   Thermal power plant of Piombino (Tuscany) under construc-
          tion.

Air and Water Alteration

      Concerning environmental alterations introduced by cooling
systems, we imagined that thermal power plants were water-cooled.
As it happens, today nearly all of them in Italy are water-cooled,
and consequently, there are no alterations in the physical conditions
of the atmosphere. Further, we considered that in water-cooled
power plants local effects of the water temperature alteration are
today reduced to such levels that they do not cause damage to the
aquatic environment. On the other hand, the availability of cooling
waters highly restricts the possibilities of power plant siting;
that implies the use of valuable land like the banks of large rivers
and seacoasts which are often densely populated or exploited as
touristic resort areas and when these areas happen to be completely
wild for this very reason are likely to be protected as a natural
wildlife sanctuary.

      In other words this implies the use of more and more valuable
land, that is more expensive environmental resource, which was taken
into account in our assessment. Additionally, while on the seacoasts
the availability of cooling water is practically unlimited, this
does not occur on the rivers. Therefore, we deemed it right to
quantify this type of resource in such a way as to make equivalent
the cost of plant installation on the rivers near the load, and the

plant installation on the sea, far from the load: this was achieved
by adding to river plant production cost the extra costs for energy
transmission, from the closest seacoast, as a compensation cost for
engaging the cooling water resource.

Minor disturbances like noise and radiointerference due to
overhead lines were for the moment disregarded.

Global Alteration: The Problem of CO_2

As possible source of global alterations, at least CO_2 emission
should be considered, since it appears the most controversial topic
in the debate over the global balance of our biosphere.

The consequences of an excessive CO_2 concentration in the atmosphere are questionable and questioned; anyway, a conventional penalization per ton of CO_2 discharged into the atmosphere might be a tool
for rationally limiting such discharge.

In our first approach analysis, however, we overlook CO_2
penalization.

The Influence of Environmental Cost

In fig. 9 the previously examined diagram is completed by taking
into account environmental cost. We added to environmental costs
relevant to electricity generation, transmission, and distribution,
the costs to be charged to emissions in fuel refining. Also, the
environmental damages linked with extraction and transport of crude
oil should be considered. Unfortunately, at the present stage of
our study we did not have sufficient elements for an evaluation,
though conventional, of such costs.

Anyhow, the addition of environmental costs does not substantially alter the diagram, which seems to show a poor environmental
component in the electric energy cost, quite disproportionate in
view of the actual difficulties encountered in electric power plant
siting.

Sulfur Oxide Emission

In order to check the validity of our approach, we report in
fig. 10a the trend of the environmental cost assessment for sulphur
oxide emission.

E   DRIFTING
R   REFINING & TRANSPORTATION
G   ELECTRICITY GENERATION
T   TRANSMISSION
D   DISTRIBUTION

▨   ENVIRONMENT
▨   ENERGY RESOURCE
☐   LABOUR
■   RAW MATERIAL

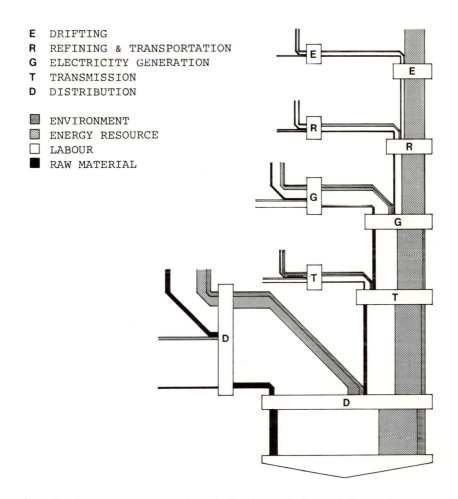

Fig. 9.   Basic resources content of electricity with environmental
          cost.

The graph exhibits the cost of pollution control and the cost
of environmental damage, assessed with the criteria so far proposed,
as a function of the degree of SO_x pollution at ground. The total
of the two costs with its minimum determines the most appropriate
compromise between pollution control cost and environment deterioration. The pollution degree x has been adopted in the economic

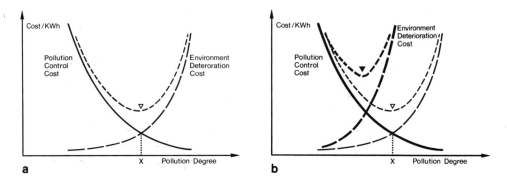

Fig. 10.

calculations of fig. 9. This pollution degree is considered as
acceptable by the present Italian laws and the relevant environmental
cost approximately corresponds to the above mentioned minimum of the
total of the two costs.

Now, let's suppose that the cost (for kWh produced) which the
community must bear for sulphur oxide pollution is in fact, due to a
reasonable mistake in its assessment, five times higher. In this
case (fig. 10b), the optimum pollution degree decreases to lower
values. This implies more expensive control measures which would
engage after all more labor and raw materials rather than environment.

Visual Nuisance

The choice of economic value of landscape nuisance is much more
questionable. A significant increase of such values does not seem
to correspond to present requirements. Indeed, the impression that
the penalization is insufficient derives from a poor correspondence
between the difficulties in building plants, namely power plants,
and the penalization of environmental disturbances. An increase in
environmental penalizations would affect more overhead lines than
the power plant; on the contrary the major difficulties are concentrated on power plant siting.

On the other hand, no doubt that the most evident element of
the power plant is the stack. Our evaluation criterion assigns to
the stack about three quarters of the total power plant visual
impact.

A reduction in stack height would drastically reduce the impact on the landscape, but increase effluents pollution. Now, stacks tend to increase rather than diminish in height and this could not occur if today the nuisance to the landscape were considered as extremely disturbing.

We could, therefore, think that a significant contribution to environment cost derives from factors which we neglected, i.e. from the environmental damages originating from extraction and handling of oil, or, but this is less credible, from CO_2 emission.

At any rate, we wanted to plot the diagram of resource mix also for an environmental unit cost increased 5 times (fig. 11).

We are thus in an extreme situation, at least by present standards, for considering the environment resources. Despite this fact, the environment resource used is still modest if compared with other resources and does not seem to represent the fundamental cost for electric power production from fuel oil.

This is strangely in contrast with the opposition of the public opinion which actually conditions the development of power plants and which largely aggravate the energy crisis. This leads us to think that this opposition, which is not so strong against other more evident and irreversible land uses is more linked with social relation problems than with actual land optimization ones.

At any rate, this type of analysis can serve, in my opinion, to clarify the aspects of this problem so important for our society.

At this stage, it would be interesting to effect the same analysis for different energy sources and to assess the advantages and disadvantages of the use of different energy sources, or of different ways of using energy, in view of different unit costs for environment and energy resources.

We do not yet have sufficient elements to do this, particularly as regards the environment resource.

In the next lecture, we will explore this topic with reference only to the cost of the energy resource, neglecting the problems related to the consumption of the environment resource.

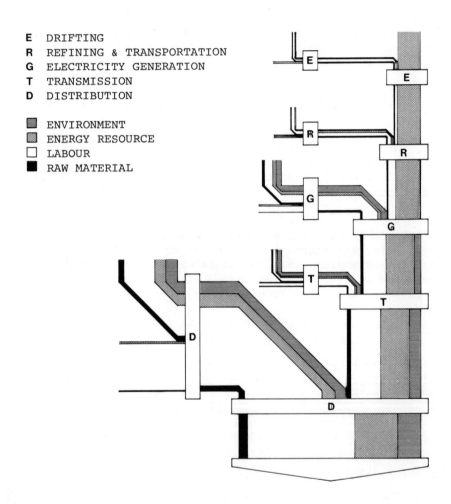

Fig. 11. Basic resources content of electricity with environmental
          cost 5 times higher.

3. COMPETITIVENESS COMPARISONS AMONG ENERGY TECHNOLOGIES RELATED
    TO ELECTRICITY

Introduction

     In the previous lecture, we made some considerations about the
validity of economic comparisons in energy choices. In particular,
we recognized the need for making such choices by assigning to the
resources a conventional money value which also accounts for the
price these resources may reach in relation to their gradual exhaustion.

     We described the difficulties of an integral application of the
above mentioned method and we supplied numerical data above all with
a view to envisaging results we would obtain by the application of
the methodology under study. The aim of this lecture is, on the
contrary, to give the results, strictly expressed in quantitative
terms, of some evaluations made, in an extremely simplified way, on
the basis of the above mentioned method.

     The economic convenience of employing new technologies in the
energy field is analyzed by supposing that fuel oil is still at the
base of the energy economy and assuming as a variable the market
price of fuel oil, expressed in actual goods (i.e. in constant
monetary terms).

     In particular, we aim at stressing the weight of the energy
consumed for the construction of power plants, namely we aim at
analyzing the phenomenon called "energy cannibalism."

     In order to simplify this analysis, my exposition will be lack-
ing in contents.

     Moreover, I thought this study could give a contribution, even
if a modest one, to the formation of a set of actual reference data,
which are today so necessary to the economic assessments of new
technologies.

     The main cost items considered in competitiveness comparisons
are listed in opposite tables, in order to give more weight to their
reference value. All costs are referred to 1975.

     The competitiveness comparisons will be made as a function of a
parameter "e" representing the unit price of fuel oil[1] in constant
monetary terms, using the 1975 average price as the unit. In other

_____

[1]Including the value of both the resource (crude oil), as defined in
the previous lecture, and all the resources necessary to oil extraction, transportation, and refining.

words, "e" = 2 means that the price of fuel oil expressed in actual goods is twice that of 1975.

Today (September 1979) the value of "e" is equal to approx. 1.2, while at the beginning of 1973, prior to the energy crisis, it was worth about 0.25 (fig. 1). The analysis considers "e" values up to 10, even if an economy based on such high-cost primary sources would probably be too different from the present one for the extrapolations to be considered as valid. In fact, the average share of primary sources in the costs of the unit of product would be in the order of 40% (for "e" = 1 such share is in the range of 7% and for "e" = 0.25 less than 2%).

Figure 2 shows the trend, as a function of "e," of the cost of the basic kWh[2] produced through thermal power plants, using as unit the cost of the kWh corresponding to "e" = 1 (1975 prices). This trend takes into account the fact that not only fuel costs (curve a) vary with "e," but even capital costs do (curve b represents the total cost of the kWh). In effect, the power plant cost increases as the cost of the energy needed to build it increases (for "e" = 10 the power plant costs approximately 40% more than for "e" = 1). Nonetheless, the capital cost share becomes smaller and smaller as "e" increases, dropping from about 30% for "e" = 1 to about 4% for "e" = 10.

It could also be advisable to measure the competitiveness of auxiliary sources with nuclear production facilities.

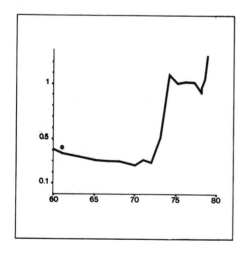

Fig. 1.

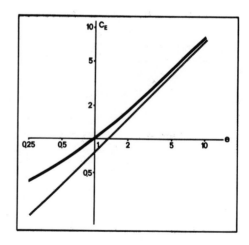

Fig. 2.

---

[2]Utilization duration 6,000 hrs./year.

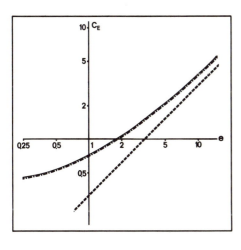

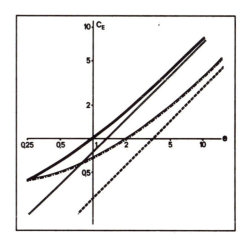

Fig. 3.   Fig. 4.

Figure 3 illustrates the cost trend of the kWh produced with lightwater nuclear power plants as a function of parameter "e_n," i.e. of the nuclear fuel cost expressed in actual goods and relative terms.

Curve c represents the share of the nuclear fuel cost and curve d the total cost; hypothesizing that the nuclear fuel cost remains proportionate to the oil cost, $e = e_n$ and the two diagrams of Figures 2 and 3 are superimposable and costs comparable (fig. 4).

All following assessments are aimed at providing indications on the competitiveness of the techniques being studied.

Cost of Electric Power from Geothermal Power Plants

The geothermal power derived from vapor-dominated hydrothermal systems is already useable today in economic terms. Instead, the economic exploitation of water-dominated hydrothermal systems, which are much more frequent in nature, is highly conditioned by the salt content of the fluids found, which may make their utilization difficult and effluent disposal problematic.

Since a significant share of the cost of geothermal power derives from exploration and harnessing activities, whose cost is highly aleatory, it is possible to express the costs of electric power from geothermal power plants only with a wide range of values.

In the calculation of the cost of the geothermal kWh, capital costs will be considered as the sum of two components, one of which includes the installation costs of the plant proper and the other the installation costs of the geothermal field (soundings, drillings, preparation of wells and pipelines to the power plant).

The installation costs of the geothermal field will be different according to the geological peculiarities of the geothermal system, depths and diameters of wells, characteristics of fluids, logistic situation, and several other factors.

In Table 1 there are indicated the extreme values expected in Italy and utilized to assess the cost of the geothermal kWh.

Fig. 5 indicates production cost estimates of geothermal energy that can still be found in Italy by means of traditional techniques and at average depths (not exceeding 3000 m).

As can be seen, geothermal electric power is competitive with thermal power plants when value of "e" is between 0.8 and 2.2.

The increase, which the cost of the kWh from geothermal power plants undergoes as "e" increases, is due to the fact that energy

TABLE 1:   INSTALLATION COST OF THE GEOTHERMAL FIELD

INSTALLATION COST	$300 \frac{\$ (75)}{kW}$	$1400 \frac{\$ (75)}{kW}$
SURFACE EXPLORATION COST	$18 \frac{\$ (75)}{kW}$	$18 \frac{\$ (75)}{kW}$
WELL DEPTH	1000 m	2500 m
DEEP EXPLORATION COST (DRILLING)	$280 \frac{\$ (75)}{kW}$	$500 \frac{\$ (75)}{kW}$
RESERVOIR ENGINEERING COST	$18 \frac{\$ (75)}{kW}$	$18 \frac{\$ (75)}{kW}$
AVERAGE CAPACITY OF A WELL	2 MW	2 MW
STEAM PIPELINES COST	$180 \frac{\$ (75)}{m}$	$180 \frac{\$ (75)}{m}$
OTHER COSTS (STEAM SEPARATORS SCRUBBERS SILENCERS, ETC.)	————	$11 \frac{\$ (75)}{kW}$
REINJECTION WELLS	————	1 WELL FOR TWO PRODUCING WELLS
SUCCESS RATIO	0.7	0.6

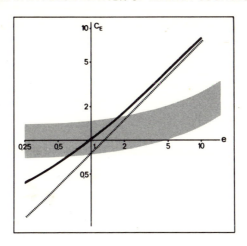

Fig. 5.

must be consumed to build and maintain the power plant. As a result, the costs increase as the cost of the primary source increases.

It is necessary to stress that, in these evaluations, we did not assign any value to the resource, as if it were inexhaustible or in any case a renewable resource. Actually, geothermal energy is not inexhaustible and it is a renewable resource only for a small part; in view of this fact, we should in effect assign it a given value. This would imply raising competitiveness toward a higher value of "e."

Cost of Electric Power from Solar Thermal Central Receiver Power Plants

As is known within the framework of thermodynamic conversion of solar energy for the production of electricity, central receiver power plants are among the plants which lend themselves best to offer significant capacities thanks also to the relatively high concentrations which can be achieved and to the possible scale economies in a mass production of heliostats.

In our calculations, we will refer to a power plant having a capacity of 10 MW connected to the network and therefore without storage units. In Table 2, there are indicated the main elements we considered to assess the cost of the kWh produced by such a plant. In particular, we considered a mirror field composed of 31 m^2 heliostats whose cost on site was estimated about 3800 $ (75), corresponding to a specific cost of 570 $ $\frac{(75)}{kW}$. This cost can be obtained only in mass production.

TABLE 2:   INSTALLATION COST OF A SOLAR THERMAL
CENTRAL RECEIVER POWER PLANT

INSTALLATION COST =   1470 $\frac{\$\ (75)}{kW}$

HELIOSTAT SURFACE	4.6 $\frac{m^2\ OF\ HEL}{KW}$	31 m^2 PER HELIOSTAT
LAND USE	2.6 $\frac{m^2\ OF\ LAND}{KW}$	172 m^2 OF LAND PER HELIOSTAT
LAND SETTLMENT AND CIVIL WORKS COST	8 $\frac{\$\ (75)}{m^2\ OF\ LAND}$	1400 $\$$ (75) PER HELIOSTAT
HELIOSTAT COST (INCLUDING SUPPORT AND CONTROL SYSTEM)	125 $\frac{\$(75)}{m^2\ of\ HEL}$ 570 $\frac{\$\ (75)}{kW}$	3800 $\$$ (75) PER HELIOSTAT
RECEIVER AND CONCRETE TOWER COST	170 $\frac{\$\ (75)}{kW}$	————
ENERGY CONVERSION COST (MECHANICAL AND ELECTRIC MACHINERY, COOLING AND AUXILIARIES)	450 $\frac{\$\ (75)}{kW}$	————

Fig. 6 shows the costs of electric power produced with plants
of this type provided that the amounts of electric power generated
be such as to enable large-scale implementation of heliostats. Also
in this case, costs increase with "e" as a result of "power" consumption needed to build the power plant.

To account for approximations in estimating the costs of the
solar power plant components, a belt of values was considered, which
is included between the value found (which is still regarded as a
not easily achievable target) and a 30% higher value.

It can be noted that solar power plants are competitive with
thermal plants only when oil costs are 5-7 times higher (in terms
relative to other goods) than the 1975 costs.

As a consequence, solar power plants of this type, at least in
our climates, belong to an energy scenario very different from the
present one. In such a scenario a lot of other new technologies for
facing fuel oil shortage would already have been more conveniently
applied.

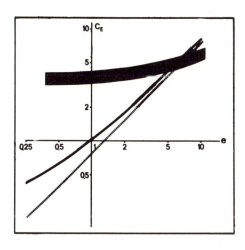

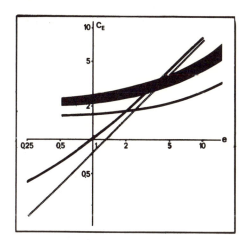

Fig. 6.   Fig. 7.

On the other hand, since no industrial country renounces to
study the thermodynamic conversion of solar energy, energy scenarios
of this type are believed possible sooner or later.

Cost of Electric Power from Photovoltaic Power Plants

In a preliminary study effected by ENEL within the framework of
Progetto Finalizzato Energetica (Energy targeted project of CNR), the
cost was assessed for a 10 MW photovoltaic power plant with mono-
crystalline silicon cells (η = 11%) without concentrators.

Table 3 summarizes the main elements we considered to carry out
this study, for which we did not take into account the cost of photovoltaic cells.

To assess the costs of photovoltaic cell module, reference was
made to the goals set by the U.S. Department of Energy for 1985 and
1990.

Fig. 7 displays the trend of the values determined above, which
represent a belt of possible costs, as a function of "e."

As regards photovoltaic power plants, we have to consider that
important conventional structures (such as structures supporting
photovoltaic modules, wiring, power conditioning, etc.) are necessary
and cannot be eliminated. At present, their cost is a considerable
share of the overall photovoltaic power plant cost.

TABLE 3:   INSTALLATION COST OF A PHOTOVOLTAIC POWER PLANT

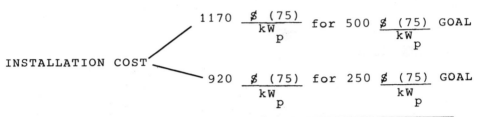

INSTALLATION COST
$1170 \; \dfrac{\$ (75)}{kW_p}$ for $500 \; \dfrac{\$ (75)}{kW_p}$ GOAL

$920 \; \dfrac{\$ (75)}{kW_p}$ for $250 \; \dfrac{\$ (75)}{kW_p}$ GOAL

MODULE (310 kW_p) SURFACE	$10 m^2 \dfrac{OF\ MODULE}{kW_p}$	$3.1 m^2$ PER MODULE
LAND USE	$35 m^2 \dfrac{OF\ LAND}{kW_p}$	$11 m^2$ PER MODULE
LAND SETTLMENT AND CIVIL WORKS COST	$4.5 \dfrac{\$ (75)}{m^2\ OF\ LAND}$	$50 \$ (75)$ PER MODULE
SUPPORT AND SETTING UP OF MODULES COST	$360 \dfrac{\$ (75)}{kW_p}$	$110 \$ (75)$ PER MODULE
OVERHEAD CONNECTIONS COST	$100 \dfrac{\$ (75)}{kW_p}$	$31 \$ (75)$ PER MODULE
CONTROL APPARATUS COST	$50 \dfrac{\$ (75)}{kWp}$	$15.5 \$ (75)$ PER MODULE
PHOTOVOLTAIC CELLS COST (DOE GOALS)	$\begin{matrix} -500 \\ -250 \end{matrix} \dfrac{\$ (75)}{kW_p}$	$\begin{matrix} -155 \\ -77 \end{matrix} \$ (75)$

To have an idea of how much the photovoltaic power plant without photovoltaic cells costs, the curve is also shown relevant to a nil cost of solar modules.

Cost of Electric Power from Wind Power Plants

We will carry out our assessments with reference to a mean-capacity wind generator having the characteristics shown in Table 4. To assess the cost of the kWh produced by wind power plants, we will refer to the available wind characteristics for some Sardinian areas where anemometric stations are located, assuming that sites are found for installation of power plants having corresponding characteristics.

TABLE 4:   INSTALLATION COST OF A WIND POWER PLANT

INSTALLATION COST   940 $\frac{\$ \; (75)}{kW}$

RATED CAPACITY	875 kW	—————
RATED WIND SPEED	12.5 m/s	—————
ROTOR DIAMETER	65 m	—————
COST OF WIND GENERATOR	820 $\frac{\$ \; (75)}{kW}$	717,500 $ (75) PER WIND GEN.
COST OF SWEPT AREA	0.3 $\frac{\$ \; (75)}{m^2}$	

TABLE 5:   DISTRICT HEATING (CHIVASSO)

INSTALLATION COST (Distribution network,heat exhangers,etc.)	$253 \cdot 10^6$
MAINTENANCE COST	583,000 $\frac{\$ \; (75)}{year}$
CAPITAL AND MAINTENANCE COST OF CONVENTIONAL BOILERS	543,000 $\frac{\$ \; (75)}{year}$
FUEL CONSUMPTION OF CONVENTIONAL BOILERS	18,000 $\frac{ton}{year}$
UTILIZATION DURATION	2,000 $\frac{hours}{year}$

Fig. 8 evidences the trends of the values determined above, which represent a wide range of possible costs, as a function of "e." As can be seen, the great variability of these costs is closely linked to the wind characteristics of the site, which determined the machine utilization.

The cost of the energy produced by wind generators should be compared with the cost of the fuel only (dashed line) required to produce electric power with thermal power plants. Indeed, the significant wind irregularity does not allow one to rely on wind generator capacity; as a result, they should be considered as "fuel savers."

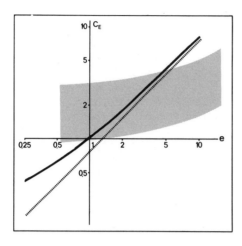
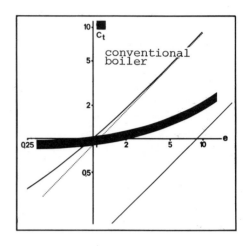

Fig. 8.   Fig. 9.

Co-Generation for District Heating from Large Power Stations
Near Towns

In principle, the obtainable energy savings with co-generation
are considerable; but the economic advantages are conditioned by
numerous local variables, such as location of thermal loads with
respect to the power plant, heating periods, heat demand, in addition,
obviously, to the price trend of primary energy sources.

As an example, we will describe a typical case with the climate
of North Italy and with power plants near consumption centers.

It is a study conducted for ENEL by a specialized firm (jointly
with the Regione Piemonte and Comune di Chivasso authorities). This
study aimed at evaluating a system for heating one part of the town
of Chivasso. For this system, we supposed to use steam bled from a
320 MW unit of the existing power plant. Table 5 exhibits the data
used in the study.

Fig. 9 shows, as a function of "e," the cost of the kcal supplied
to the consumer as against the cost of the kcal supplied by conventional boilers.

From the figure, you can realize how under the Italian conditions
(at least under the conditions similar to those given in the example),
district heating before the energy crisis was not economically convenient in comparison with conventional boilers, and this is why it
was not largely employed.

Co-Generation in Small Total Energy Systems

In addition to co-generation for district heating, the attention
has been recently focused on co-generation in small total-energy
systems. These systems can be mass-produced and, therefore, can
allow significant savings in capital costs, especially in costs relevant to engines if we resort to a series production.

One well-known project in this field is the Totem system for
combined production of power and heat, developed by FIAT.

This is a limited capacity module (15 kW, 33,000 kcal/hr),
consisting of an engine from the 127 motor car, fueled by gas and a
synchronous generator connected to it. A group of heat exchangers
recover the heat from the heat engine, exhaust gases, and generator.

Based on the data contained in Table 6, the cost of the kWh
produced by the Totem system was assessed as a function of "e"
(fig. 10).

Transmission from Remote Energy Sources

In today's search for new energy sources, which are often very
expensive, important hydroelectric resources far from consumption
centers (or resources which are not easily transferable) are disregarded as they are considered not convenient because of their considerable transmission costs. On the other hand, considering our
great efforts towards energy conservation, transmission is to be

TABLE 6:   TOTEM

INSTALLATION COST OF TOTEM	5,100 $ 75
MAINTENANCE COST OF TOTEM	24.5 $\frac{¢(75)}{h}$
INSTALLATION COST OF BOILER	1000 $ (75)
MAINTENANCE COST OF BOILER	5.1 $\frac{¢(75)}{h}$
FUEL CONSUMPTION (TOTEM)	6 $\frac{N\ m^3}{h}$
NATURAL GAS COST	8.8 $\frac{¢(75)}{m^3}$
UTILIZATION DURATION	2000 $\frac{hrs}{year}$

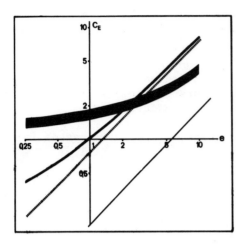 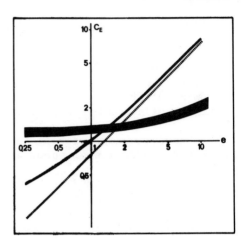

Fig. 10.   Fig. 11.

looked at as a means to recover these sources, that is it may be
regarded as an alternative source of energy. In this connection,
assumptions should be made on the cost of electric energy produced
by a remote source. We will assume as a remote source a hydroelectric
power plant producing electric energy at a cost equal to that of
electric energy produced by a thermal power plant (near consumption
centers) in the case of e = 1.

Fig. 11 shows the variation, as a function of "e," in the cost
of the kWh produced by a hydroelectric power plant, evaluated at the
end of a transmission line with a length of 1,000-3,000 km.

We can observe how this particular resource becomes competitive
before other energy sources which are today considered as alternative.

Synthesis of Competitiveness Comparisons

The figure 12 summarizes the results of the competitiveness
comparisons carried out in the previous paragraphs.

For each of the new technologies, the range of values of "e"
is given, starting from which the technology under consideration
becomes competitive.

New Utilizations

The method followed so far for competitiveness comparisons
among the various energy forms may also be applied to evaluate the

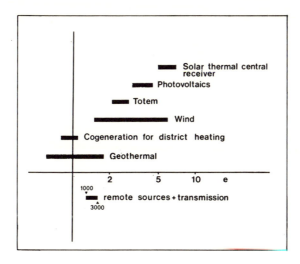

Fig. 12.

benefits deriving from new utilizations substituting the conventional
ones consuming oil. Firstly, we will take into account the sector
of road transport.

Electric energy is above all qualified to make mechanical work
available. Almost all the needs of such mechanical work are covered
today by electric power, except for transport in independent vehicles,
where only hydrocarbons are used. In particular, road vehicles rank
extremely high among oil consumers. It is thus logical that priority
be given, in the sector of users, to the "electric road vehicle."

Today, the diffusion of the electric vehicle is essentially
conditioned by the problem of power storage. The storage capacity
per unit of weight and volume of present batteries limits the range
of such vehicle significantly, narrowing its use to urban traffic
alone (range 70-100 km).

On the basis of the data supplied in Table 7 we have obtained
figure 13. By referring to this figure, one gets an idea of the
competitiveness of the electric vehicle as compared to the combustion
one (provided that the electric vehicle be produced in a number comparable to the corresponding thermal vehicle), even supposing that
the electric power is produced with fuel oil, namely that the power
comes from the same primary source. This figure indicates the ratio
between the cost of electric power for an electric vehicle and that
of fuel for an internal combustion one.

If the annual capital and maintenance costs for the two vehicles
are considered to be equal, when such ratio reaches 1, it represents

TABLE 7:   COMPARISON BETWEEN ELECTRIC
AND COMBUSTION VEHICLES

	COMBUSTION VEHICLE (Diesel)	ELECTRIC VEHICLE
PAYLOAD	1.500 kg	
ANNUAL MILEAGE	20.000 km	
CONSUMPTION	$0.2 \frac{litres}{km}$	$0.6 \frac{kWh}{km}$

the economic limit of the combustion vehicle. The equality of fixed annual costs can be an achievable goal, in that the electric vehicle should have a longer life and lower maintenance costs, to offset the high costs of batteries.

Belt "a" corresponds to the case in which day electric power is used to recharge batteries, belt "b" to the case where the batteries are recharged using to a good extent night electric power.

It can be observed that the electric vehicle is already competitive today in case "b," whereas in case "a" it becomes competitive when "e" = 2. At any rate, the advantage of the electric vehicle increases as the value of "e" increases.

ENEL has carried out tests on vehicle prototypes to be used for the urban network maintenance services for the distribution of electric energy. Two vans, derived from the 850 T, built by FIAT, have been equipped and tested on the road by ENEL for the last three years.

The result of this common effort is a type of van which, produced on a small scale, is used in a demonstration fleet which is now used in routing jobs at the ENEL division of Milan.

Another demonstration fleet of 3-5 passenger vehicles is now operating at ENEL for use in ten Italian towns for management services.

Outside the sector of transport, the only other important new users of electric power are found in the field of thermal uses. In

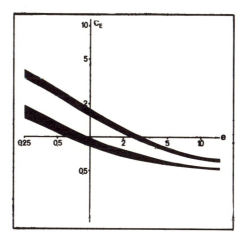

Fig. 13.

this case, equipment must be used which enables use of electric power, taking full advantage of its high-grade power characteristics. In this sector, we are now considering an organic plan of demonstration activities with regard to the sector of "heat pumps."

At present, a study is being carried on as to the possibility of using a heat pump in combination with solar energy for home heating. Demonstration will be made using a number of demonstration systems for the heating of ENEL personnel homes at the thermal power plant of Rosano Calabro.

4. ENERGY STORAGE IN THE ELECTRIC SYSTEM

Introduction

In introducing the subject of storage, I think it is interesting to call your attention to the definition of energy. The concept of energy can be expressed in many different ways, but out of all of them I choose the following one: energy is "stored mechanical work." Though slightly embarassing to talk about storing something which, if not stored, doesn't exist, this definition immediately evidences the close relationship between energy and storage.

At any rate, overlooking the misleading definitions, it is true that energy is usually considered as goods and as such it is produced, transformed, transferred, and also stored. Storage, therefore, is one of the fundamental aspects of energy handling in energetic systems.

In the following pages, I will not talk about fuels which, as actual commodities, do not raise particular storage problems. I will, instead, talk about energy storage under other forms.

As long as energy was cheap, the role of storage was limited; today, like all problems concerning energy conservation, it is a topical subject.

Within this general context old ideas are being reconsidered which now, however, can be effectively carried through thanks to the development of new technologies.

In view of the broadness of the subject, a number of limitations must be set. Consequently, I will leave out the problem of heat storage in addition to that of fuel storage, and only examine the problems posed by the storage of energies of highest value, such as electric energy and mechanical energy, which are perhaps the most difficult to store.

The end uses of these energies are mainly mechanical work, electricity for information, processes and light; rarely, we find heat as an end use of these energies. Electricity and light are mainly supplied by the electric system. The users of mechanical work can be divided into two categories: self-powered means of transport which do not use electricity and those which, instead, use such carrier and therefore are fed by the electric power system.

Therefore, we will talk about the storage of energy in the electric energy system and in the self-powered means of transport; namely, in the means which can cover distances not being bound by special continuous supply systems.

Reasons for Energy Storage

Storage fulfills the main purpose of lessening the constraints between the energy demand and availability at production level. Since it allows one to store energy when availability exceeds the need and then redistribute it in case of need, storage features numerous technical and economic advantages. We will examine the reasons for storage as a means for energy conservation, which at present is that receiving the most attention.

Now, in terms of energy conservation, four reasons can be singled out for storage, three of which stem from the fundamental considera- tion that still today fuels, whether fossil or nuclear, are by far the main source of energy. Thus, to transform such energy into mechanical and electrical work, we must use thermal prime movers.

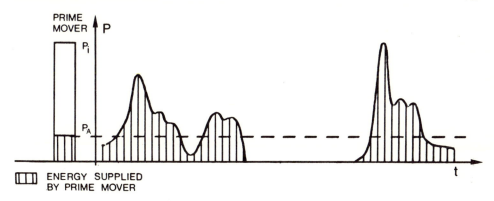

Fig. 1.

Let's consider a possible time distribution of the power requested versus time of our prime mover (load diagram)(fig. 1).

A thermal prime mover must be designed for the peak power requested. Notice that the average value of the power supplied, P_a, is much lower than the installed power, P_i, and thus if the prime mover is designed for the peak power its utilization is modest.

Now let's consider that there is a wide variety of thermal prime movers which are competitive with each other: engines with high efficiency featuring low operating costs but with a high installation cost and, vice versa, engines with low installation cost and thus with low efficiency and high operating cost.

The high efficiency engines are more competitive for long-duration utilization while the others are preferred for short-duration utilizations. In the electric system, for instance, we find a large variety of prime movers ranging from those with a high operating cost and low installation cost, such as the turbogas units, which are used to cover peaks with utilizations below 1000 hours, up to nuclear ones with low operating costs and high installation cost. The same occurs in motor vehicles: if the engine is not utilized so much, a gasoline-fueled engine is used; if it is much utilized, a diesel engine is resorted to.

Coming back to our case, we are inclined to use an engine with low efficiency with a consequent energy waste in order to reduce installation costs. The situation can be improved by using a storage system. The prime mover is thus designed for average power while the peaks are covered by the storage system which stores the energy available during low demand periods (see fig. 2). A small prime mover, having a very high efficiency, can then be used, providing considerable energetic advantages.

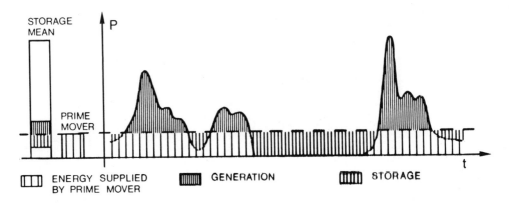

Fig. 2.

One of the fundamental applications of this principle is the hybrid car, that is to say a car with a small engine of limited power which produces energy that is stored in batteries (or in a flywheel, as we will see later on) and this stored energy covers the service peaks.

Let's analyze the second point. If we use the prime mover designed for the peak power demand and it is not utilized much, we have energy called marginal energy which remains available and which actually is cheap since its conversion is performed by a prime mover already paid for (fig. 3). In this case, we are inclined to use this energy also for improper uses.

This occurs in particular in the electric system where the marginal electric energy is made available at a reduced rate (in general at night) and is thus often used for heating purposes, an improper energetic utilization.

The pumped storage system, vice versa, allows one to upgrade the energy of off-peak hours, thus drastically reducing the availability of marginal energies.

The third point is as follows: the load diagram which was entirely positive in the previous cases, may also be negative. This means that the system may need braking (see fig. 4). This energy is generally dissipated as heat.

In such case, if we have a pumped storage system, we can store energy during the braking, thus reducing the average energy produced and gaining part of the energy that can be produced by the prime mover (fig. 5).

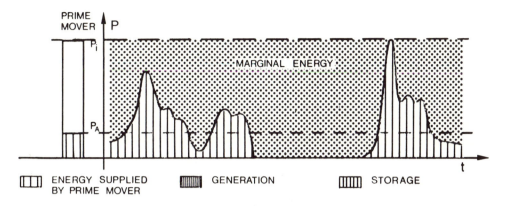

Fig. 3.

The third point is, therefore, the recovery of braking; recovery of the energy which otherwise would be wasted.

So far, we have always talked about a thermal prime mover, namely energy derived from fuels. When, instead, the energy is directly derived from available mechanical energy, such as hydroelectric energy or wind energy, the problem is set under slightly different terms.

Let's consider this relatively regular load diagram (fig. 6) of a group of electric users and the diagram which shows the availability of the primary source; this diagram is very irregular because of the

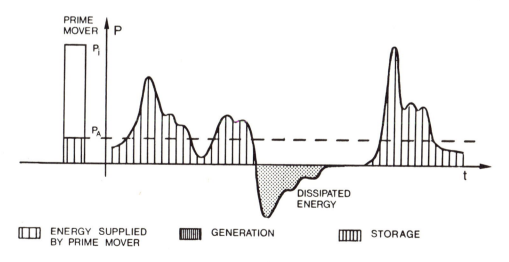

Fig. 4.

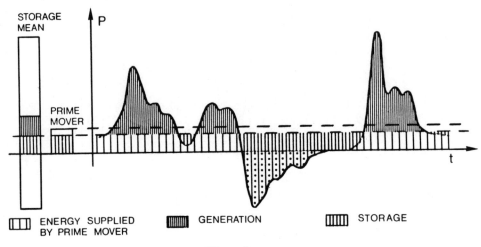

Fig. 5.

nature of the primary source.

The power availability depends on water flow (the peak may be a flood) if hydroelectric energy is involved and on wind speed in the case of wind energy. Even if the prime mover is designed to cover the peak load (see Fig. 7) but a storage system is not provided, not only is it impossible to always cover the load, but a lot of energy available ends up being wasted. By contrast, with the storage, energy is stored instead of being wasted and is released during power shortages. In this case, however, the prime mover should be designed for the availability peaks (fig. 8).

This actually occurs only when using the wind source, because wind energy cannot be directly stored but must be converted by means of a prime mover large enough to use such energy and then inserted in an electric system featuring an adequate storage capacity.

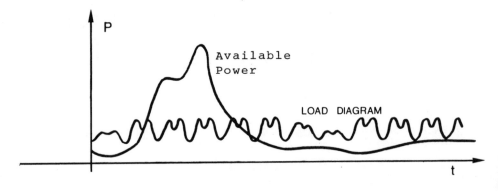

Fig. 6.

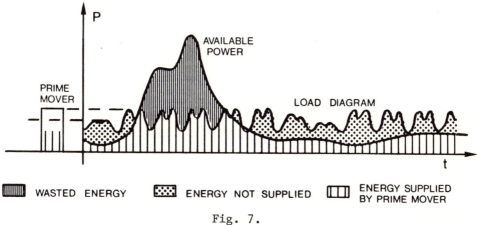

Fig. 7.

Hydraulic energy, instead, can be directly stored before its conversion. This makes it possible to use a small prime mover. This explains why the first large storage systems were of hydroelectric energy.

Based on the above, four reasons for storage can be established as far as energy conversation is concerned: 1) improvement in thermal conversion efficiency (namely of the prime mover efficiency), 2) upgrading marginal energy, 3) recovery of braking, and 4) utilization of the discontinuous primary source.

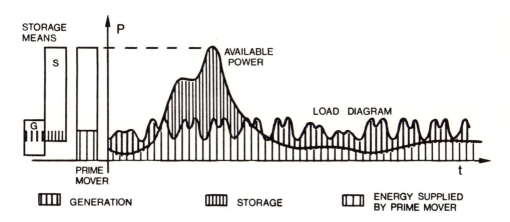

Fig. 8.

Applications of Storage in the Electric System and in Traction

In the electric system (Table 1), conventional storage is a hydroelectric power plant fed by a reservoir acting as a flow regulator. This type of storage, used since the beginning of electric systems, has been fundamentally used for utilization of discontinuous primary sources; today it is used also to improve the thermal conversion efficiency. In effect, if stored energy is used in the hydraulic form to cover the peak loads, low-efficiency prime movers like the turbogas units can be eliminated and prime movers with a higher efficiency can be better used. These functions are peculiar of pumped storage plants, that is hydroelectric plants which during off-peak hours take water from a lower basin through a system of pumps and fill the upper reservoir. The two main purposes of storage plants today are the improvement of thermal conversion and the upgrading of marginal energy.

These storage plants can be also used to improve the utilization of the discontinuous primary sources and in particular play an important role in the development of alternative sources.

In traction (Table 1), the two main applications of storage are in the electric vehicle and in the hybrid vehicle.

The hybrid vehicle, already mentioned, enables to improve thermal conversion efficiency and to allow braking recovery.

The electric vehicle, which at the beginning of the car development was greatly considered, has over the past few years become an important possible utilization, because of both environmental and energy crisis, receiving extremely widespread consent from both the technical circles concerned and public opinion. The electric vehicle

TABLE 1.   ROLE OF ENERGY STORAGE

Reasons For Storage in Energy Conservation	Application in			
	Electric System		Traction	
	Reservoir Plant	Pumped Storage	Hybrid Car	Electric Car
Improvement of Thermal Conversion Efficiency	O	●	●	●
Recovery of Excess Energy		●		●
Recovery of Braking			●	●
Utilization of Discontinuous Energy Sources	●	O		

is fully fed by energy stored (generally in batteries but the flywheel is also proposed) during the rest periods of the vehicle and
thus mainly at night. Consequently, storage in the electric vehicle
allows us to reach the following goals: improvement in thermal conversion efficiency, valuable recovery of excess energy, and lastly,
braking recovery.

I would like to draw your attention to these three goals since
I feel that, in the present energy situation, they justify the re-
newed interest in the electric vehicle and may be the impetus to overcome the major obstacles to their diffusion. The most important of
such obstacles is the excessive weight of the batteries per unit of
energy that can be stored, which drastically limits the energy storage
capacity and thus the range of such a vehicle.

Storage Units and State of the Art of their Application

The types of qualified energy which are stored are essentially
two: mechanical energy and electrical energy (Table 2).

Mechanical energy can be kinetic, elastic and due to the force
of gravity. These forms of energy storage have been used by man for
a long time: the mechanical energy due to the force of gravity and
the mechanical energy of the elastic type were used to run clocks
(e.g. weight-driver or spring-loaded clocks), while kinetic and elastic energy represented the first instruments of man. In effect, if
you think of the club, it merely stored kinetic energy and released
it shortly after when used in hitting. The club then evolved into
the hammer and into other percussion tools and then into more complex
machines, such as the eccentric press, which is one of the applications of energy storage in the kinetic form of the flywheel.

TABLE 2. MAIN APPLICATIONS OF STORAGE

Energy Type		Storage Unit	Electric System	Traction
Mechanical Energy	Due to Gravity	Hydro Reservoir	●	
	Kinetic	Fly Wheel	▲	■
	Elastic	Compressed Air Reservoir	■	
Electric Energy	Electro Chemical	Battery	■	●
	Electro Magnetic	Superconducting Magnet	▲	

● Operating   ■ Experimental   ▲ Under Study

The first application of storage of elastic deformation energy was certainly the bow.

All these applications exploit a capability typical of kinetic energy storage, i.e. the possibility of making the energy stored available in the very short time, i.e. of providing considerable capacity peaks.

Electric energy can be divided into electrochemical and electro-magnetic energy. The electromagnetic storage of considerable amounts of energy has been considered only since the development of the tech-nology of the superconductors. These allow one to obtain very intense magnetic fields practically with very low losses. The electrostatic storage, on the contrary, is not feasible, since the material with super-insulating properties has not yet been invented.

Let's examine now how the different forms of storage are applied in the electric system and in traction.

Fig. 9 shows the general arrangement of Lake Delio pumped storage hydroelectric plant in Northern Italy. It uses a head of 700 meters between the upper reservoir of Lake Delio and the lower basin of Lake Maggiore. Generation power is about 1,000 MW with around 16 million kWh of annual energy output. This plant, in operation since 1974, has been the first major project in the extensive pumped storage program which ENEL, taking advantage of Italy's favorable orography, is implementing (8,000 MW capacity plants are at present under con-struction) so as to achieve an in-depth regulation of electric power supply on the Italian network. Fig. 10 shows a view of Brasimone pumped storage plant, a 300 MW plant in operation since 1977.

Fig. 9.   Artist's view of pumped storage plant of Lake Delio.   On the left lower corner is the underground powerhouse.

Fig. 10.   Brasimone-Suviana pumped storage plant.   Powerhouse is
           underwater to a depth of 45 m.

        Extensive programs of research, development, and demonstration
are underway in the field of battery-powered vehicles, as the widespread use of electric vehicles (fig. 11) should have a positive
effect both as far as energy sources diversification and pollution
reduction are concerned.   Moreover, overnight battery recharging
would contributed to the levelling of the electric load profile.

        As for storage means presently under test for traction, we must
quote the flywheel, which thanks to the progress in material

Fig. 11.   FIAT-ENEL 900 T Electric Van belonging to an experimental
           fleet used for activities connected with construction and
           maintenance of urban distribution networks: maximum
           speed is 55 km/hr, urban range of 50 km.

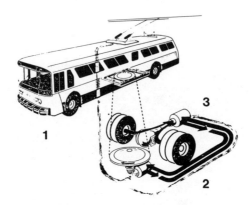

Fig. 12.   Design of S. Francisco electric trolley-bus:
          (1) Trolley draws current from overhead lines to run both
               the flywheel motor and the drive motor.
          (2) With trolley retracted, spinning flywheel's stored
               energy generates current to run drive motor.
          (3) Going downhill with trolley retracted, drive motor
               becomes generator driven by wheels, providing current
               to increase flywheel spin.

technology features a density of storable energy higher than batteries.
Since the storable energy is directly proportional to the strength of
the flywheel, the task is to succeed in finding extra light, reliable
materials with a high strength so as to realize the so-called super-
flywheels. Some new plastic fibers have the same density as aluminum
and an extremely high strength, much higher than steel. These fibers
will allow manufacture of flywheels with extremely high storage
amounts per unit weight. The flywheel has thus wide prospects in
traction. Fig. 12 shows a trolley bus designed some years ago in
San Francisco. This bus stores energy by means of the flywheel during
the route section where it is fed by an overhead line; then by utili-
zing the energy stored, it can cross downtown areas, without needing
to contact lines, for a range of 10-15 km.

    In the electric system, the reservoir for compressed air storage
is being tested. Turbogas units are being developed, which have a
compressor separated from the engine; at night, air is compressed
and injected into the turbogas unit (fig. 13) which does not drive
the compressor and, therefore, has a much higher power and provides
a greater amount of energy. The compressed air storage plant of
Huntorf (West Germany) utilizes a reservoir excavated in salt caverns,
while a similar Italian project will utilize the empty geothermal
reservoir near Sesta (Tuscany). The empty reservoir of Sesta has a
volume of about 3 million cubic meters, filled with CO_2 at 20

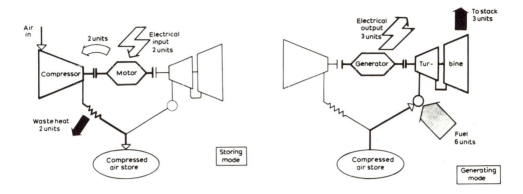

Fig. 13.   Schematic flowsheet of a compressed air storage system for
electric utility load leveling. The compressor is driven
by nighttime, off-peak electricity and the air is cooled
before being stored in an underground cavern (left). On
release, the compressed air allows fuel to be burnt twice
as effectively as in the simple turbo generator powering
its own compressor.

atmosphere pressure. Feasibility studies are underway, while the
experimental plant, to be built in 1982, shall feature 40 MW of
output power.

     Finally, even in the electric system operation there is a
tendency to resort to batteries as storage units. For applications
of this type, batteries are required which have a cost (here weight
is no longer important) sufficiently low to replace the pumped
storage plant. Since, contrary to what happens for pumped storage
plants, the overall power of a battery storage system can be fractionated in many units of reduced power, the single storage units
can be located at the high voltage transformer rooms. In this case,
the batteries can even serve as "reserve standby" for the entire
primary transmission and distribution system. Fig. 14 displays an
artist's view of a battery system which Westinghouse expects to
market in a couple of years at prices even more competitive than the
pumped storage plant.

     Still for the electric system, studies are underway on the fly-
wheel and on the supermagnets, i.e. superconductive magnets to be
located underground. Fig. 15 shows a flywheel system for energy
storage; the rotating mass is buried underground for safety reasons,
even if the adoption of fiber composites minimize the problem of
fragment containment in case of wheel failure. Tests were conducted
to study the behavior of fibers in case of failure and it was seen
that when fiber composites are excessively stressed, they crumble

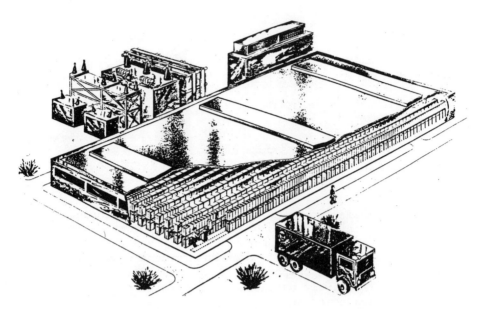

Fig. 14.   Electric storage through battery system according to a
           Westinghouse project.

or pulverize, instead of breaking into large pieces, as happens with
steel flywheels.

     Superconducting magnetic energy storage systems (SMES)   store
electric energy in a magnetic field produced by circulating current
in the winding of a magnet.   In principle, d.c. current induced in
a superconducting winding will flow without electrical losses until
energy stored is returned to the network.

     For a utility size SMES, feasibility studies performed until now
mainly in the U.S. consider plants located several hundred feet below
ground in solid bedrock (fig. 16).   This large magnet, made with
several superconducting coils is contained in a cryogenic envelope
with vacuum insulation.   A liquid helium flow at 4.2 k, or even better
a superfluid helium flow at 1.8 k, at low pressure will keep coils in
the superconducting state.

     Typical evaluated dimensions of an underground tunnel housing a
10,000 MWh SMES plant are about 7 m wide, 100 m in height and 150 m
in radius.   Magnetic forces on coils would be transmitted to the
surrounding rocks through solid thermal insulators.   Some U.S. reports
on this field conclude that these systems should be feasible even with
present-day technology.   Declared technical maximum efficiency is
about 90%, although optimization studies show that lowest annual

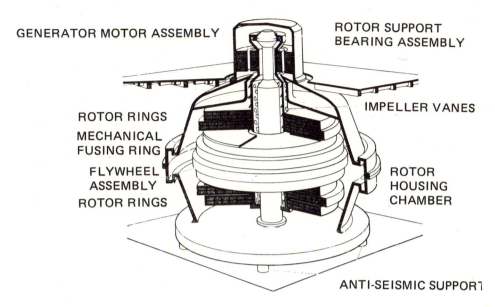

Fig. 15.   Design of superflywheel for the coverage of electrical
peak load (from <u>Scientific American</u>, December, 1973).

Fig. 16.   One of the poles of the superconducting magnet designed
for the 3.7 m European bubble chamber at CERN. Operating
at liquid helium temperatures, the magnet stores less than
one quarter of a megawatt-hour. Much larger magnets
would be necessary to provide bulk electrical storage at
economic cost.

owning and operating cost should be got for efficiencies of the order
of 80-85%.

Storage System Design

Now, I would like to dwell on the criteria which should be
followed in designing storage systems.

First of all, let's analyze the characteristics to be designed
in a storage system and, to this effect, we refer to the two diagrams
of Fig. 17. The first plots the power required and shows the capacities which the system should offer in terms both of pumping and generation; the second plots the integral indicating the energy stored
in the reservoir and shows the size of reservoir E. With reference
to a pumped storage system as a typical storage plant, we define as
pumping capacity the maximum capacity with which the energy to be
stored is consumed and as generating capacity the maximum capacity
with which the energy is released by the storage system and as reservoir the part of the system where energy is actually stored (e.g.
flywheel, battery, etc.). As you can see in the two diagrams, energy
in the reservoir decreases when the system supplies energy and increases when pumped storage operates. Therefore, the characteristics
to be designed in the plant are: the generating capacity, the pumping capacity, and the size of the reservoir. Fig. 18 shows in the
upper diagram the effects deriving from an insufficient designing of
generating capacity: the power demand is not entirely covered at
peak-time. Fig. 18 in the lower diagram shows the effect of an insufficient reservoir size: the power demand is not covered at the
time when the reservoir becomes empty. The effects of an insufficient
designing of the pumping capacity is indicated in the diagram with
the drop in capacity shown in black in Fig. 19: the slope of the
recovery curve becomes lower with consequences similar to those of
an insufficient designing of the reservoir: impossibility to meet
the demand due to emptying of the reservoir.

Consequently, we can have two types of outages: one due to lack
of power and one due to lack of energy in the reservoir. A generation deficit brings about a deficit due to lack of power, while both
an insufficient reservoir capacity and an insufficient pumping capacity bring about an outage due to lack of energy.

The problem of designing would be simple if the load diagram
were defined in a deterministic way; actually these diagrams have
always a high random component. Just think of the load diagram of
an engine driving a vehicle which strictly depends on route and traffic conditions. As a result, we cannot eliminate in an absolute way
either the lack of power or the lack of energy; we should content
ourselves with reducing the risk that these events occur to reasonable
levels.

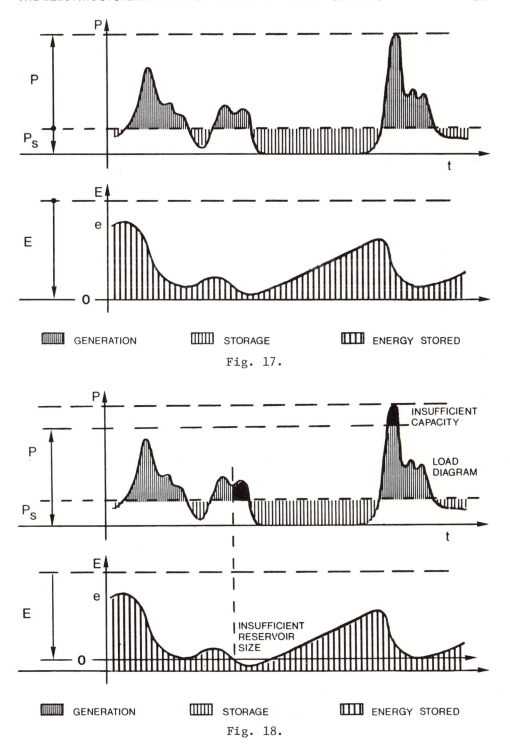

Fig. 17.

Fig. 18.

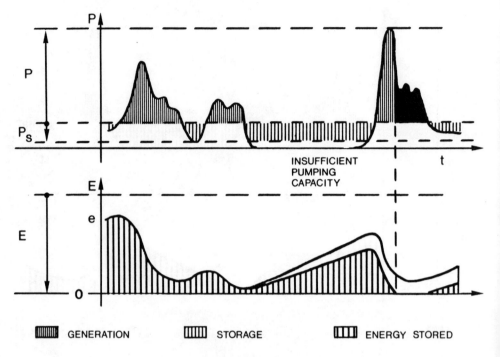

Fig. 19.

The three basic dimensions, generating capacity, storage capacity, storable energy, can be expressed in relative terms for the sake of simplicity. Assuming the generating capacity as 1, we identify the pumping capacity expressed in p.u. of the generating capacity with and the ratio between storable energy and generating capacity with h (see Table III).

We will thus have the relative magnitudes π, restoring ratio and h, which represents the number of hours when the power plant can operate at full capacity starting from the full reservoir state and reaching the empty reservoir one.

I will deal in particular with designing of the reservoir, i.e. of the choice of the optimum value of h.

As we already observed, there is always the risk of not being able to meet the demand due to lack of stored energy; the greater the reservoir, the smaller the risk. If we manage to quantify the risk (and, as we will see, this possibility exists), we will obtain a curve. This curve indicates that the risk decreases when the reservoir size (see fig. 20) increases. But, the reservoir cannot be enlarged indefinitely for evident cost reasons. Indeed, the installation cost of the plant grows with the reservoir size; therefore,

TABLE 3.   BASIC DIMENSIONS OF A STORAGE PLANT

		Per Unit of Generating Capacity
Generating Capacity	P	1
Storing Capacity	P_s	$\pi = \dfrac{P_s}{P}$
Storage Energy	E	$h = \dfrac{E}{P}$
π Restoring Ratio h Generating Hours at Full Capacity		

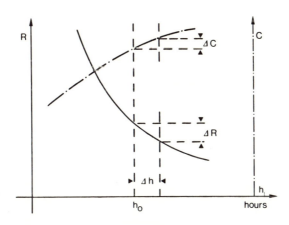

Fig. 20.

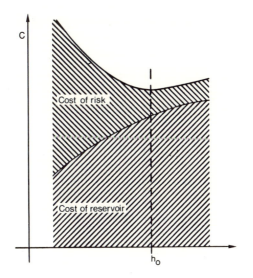

Fig. 21.

designing h, is a matter of compromise between risk and cost. We
will then say the value of h is optimum when by increasing it by a
given quantity   h, the resulting risk reduction   R might, in our
opinion, justify the corresponding cost increment. This implicitly
introduces the notion of "risk cost," that is the economic assessment
of an aleatory event; this is currently made and represents the insurance premium required for remunerating a random event.

     This notion becomes so necessary that it should be explicitly
introduced by economically penalizing the risk. In this case, the
compromise between cost and risk can be obtained in the most objective and reliable way, as an economic optimum, by summing up the risk
cost (see fig. 21) and the installation cost, and minimizing this
sum.

     We will limit ourselves to risk evaluation in the case of a
storage plant which is part of an electric system. In this instance,
we should bear in mind that the load can be known and forecasted with
sufficient reliability, so much so that aleatory components can be
neglected.

     By contrast, in a system prevalently made up by a highly forced
unavailability rate, the total available capacity is highly aleatory.

     Fig. 22(a) shows the load diagram of a work day from 6 to 6 of
the following morning and a possible mix of the generating units
meant to cover this load; the total capacity of the generating units
is greater than the peak load of a certain quantity called reserve;
this mainly is meant to cover forced or planned unavailabilities of
the various units.

     Now, let's imagine that on the day considered, the availability
is not total. For reasons of operating cost, we will use all nuclear

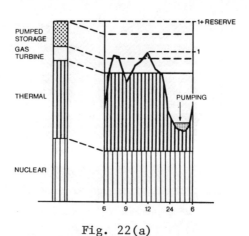

Fig. 22(a)

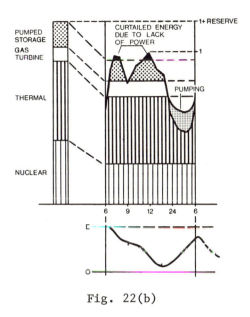

Fig. 22(b)

power plants, then all thermal power plants and, if necessary, the
turbogas units and the storage plants as illustrated in this figure.
In this case, only part of the storage plants are required to cover
the load.

Certainly, this system is a most reliable one since, with a
short duration utilization of storage plants, it avoids the risk of
reservoir emptying. But, it is not necessarily the most economical
one. Indeed, in spite of the relatively low efficiency of the storage
plant (approximately 70%), the energy produced by the turbogas units
may cost more than the off-peak energy used to pump water into the
upper reservoir.

Therefore, it could be useful sometimes to revert the position
of turbogas units and of pumped storage units; pumped storage units
are used in a more extensive way, while the turbogas units are limited to peak-load (consequently, all of them are used as reserve). In
this case, the storage performs the function of energy transfer from
hours of low marginal cost to hours of high marginal cost, which
allows fuel and thus energy savings.

From the standpoint of availability, there may be good days,
like the ones considered so far, and critical days (see fig. 22(b))
when the available capacity is not sufficient to cover the load. In
this case, part of the power demand will not be covered due to lack
of power (blackened areas in the figure).

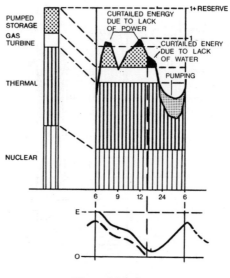

Fig. 22(c)

As concerns storage plants, it is necessary, in such critical situations, to work in such a way as to obtain the maximum reliabilit i.e. maximum utilization of turbogas units (possibly even to pump) before resorting to storage plants. In this way, the risk of reservoir emptying, and thus of not covering the demand due to lack of power is avoided.

In effect, the behavior of the reservoir level indicated in the lower part of Fig. 22(b) shows that the reservoir did not become empty. But, if the critical situation continues even on the following day (fig. 22(c), the storage plant will have the reservoir empty before finishing its service (dotted line in the reservoir level curve). Consequently, it will be unable to supply energy to the load due to lack of stored energy.

If we simulate several times the operation of the system for the entire year we are interested in, and each time we take note of the energy not supplied due to lack of power and of the energy not supplied due to lack of energy, and we determine their averages, we obtain values which tend to stabilize and give an indication of the system capability to perform the service required.

This simulation can be made with an appropriate computer program which, for each hour of the year, based on stochastic models of power plant availability, determines which power plants will be out of service at that time and how long they will remain out of service, and then assess, at the same time, the available capacity. Then the program examines if and to what extent it is necessary to resort to the storage power plants to cover the load or if, vice-versa, it is possible to store energy and how much. In this way, it will be possible to know, starting from data known at the beginning of the hour, the energy stored in the reservoir of the storage plant at the end of the hour. Proceeding in this way, we succeed in simulating the behavior of the energy stored in the reservoir during the year and particularly when the energy in the reservoir is short at the time when it is demanded.

Obviously, such simulation should be repeated for a considerable number of years, so as to achieve average results which are as close as possible to the expected values. The diagram of fig. 23 illustrates the sequence of the energy not supplied due to lack of available capacity and to lack of stored energy, values obtained in about one hundred simulations of the same operation. It is worthwhile to underline the different natures of the two phenomena: while the lack of power determines relatively modest crises almost every year, the lack of stored energy determines much rarer but more severe crises.

In this connection, it is interesting to compare two random samples of this set of values (see fig. 23(b)). The two samples might well represent the experience of two operators, A and B, who

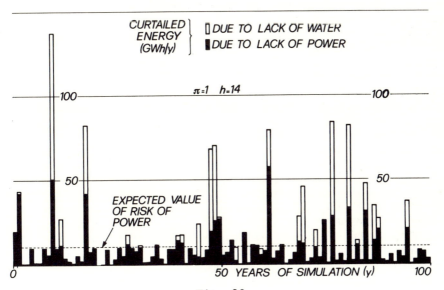

Fig. 23.

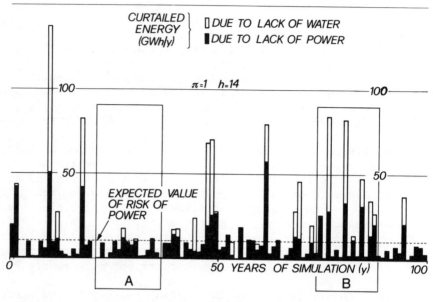

Fig. 23(b).

have managed the same system for about 15 years; their different experience would be due to purely random facts. If the two operators
met after this experience, operator A might state that outages from
lack of stored energy are not a problem, while operator B, severely
hit, will assure that the reservoir of his pumped storage plants are
entirely insufficient.

This proves that when random facts dominate, the operating
experience, even if long, cannot be considered as a design rule without a thorough interpretation, based on simulation tools.

The average value of energies not supplied, obtained from a very
high simulation number represents instead a very valid index.

Fig. 24(a) displays the behavior of the two indices concerning
outages due to lack of power and to lack of energy as a function of
the value of h (generation hours at full capacity). For h = 5 hours,
the ratio between energy not supplied and energy supplied is about
$5 \cdot 10^{-4}$, as if for one hour every year all the capacity of the system were not supplied. Sometimes, speaking about storage plants, a
discrimination is made between 5-hour "daily" plants and 13-hour
"weekly" ones (points d and w of fig. 24).

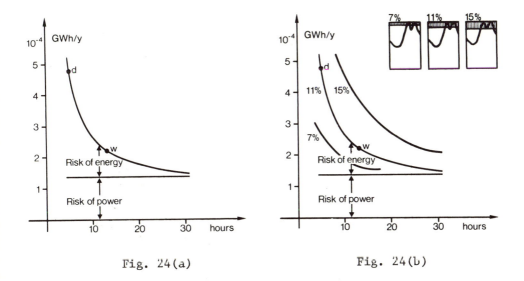

Fig. 24(a)   Fig. 24(b)

Obviously, this discrimination is absurd; designing of the reservoir is not related to the cycle duration but to the entire designing of the generating mix and to the risk of outage due to lack of water which we are willing to accept.

Fig. 24(b) shows that this risk depends also on how many pumped storage plants we want to introduce into the generating mix; indeed, with reference to a generating mix including 11% storage units, larger reservoirs are required to have the same risk in a generating mix with 15% storage units. If we want to equip the generating mix with 7% storage units, the reservoirs can be smaller. This is evident, since if we decrease the percentage of pumped storage plants, we affect less and less significant parts of the load diagram. But, the optimum size of the reservoir clearly depends also on the cost of the storage plant and, in particular, on the cost of the reservoir. In other terms, the unit cost of a pumped storage plant can be expressed as a function of the reservoir size h. Fig. 25(a) exhibits the cost functions of three storage plants with different characteristics: a high-head pumped storage plant, a mean-head pumped storage plant, and an electric battery system.

As regards pumped storage plants, the power cost depends on the water supply system cost, whereas the energy cost depends on the two upper and lower reservoirs and, in particular, on their size.

In high-head plants, the expense for the two reservoirs is generally modest with respect to the cost of water supply systems and, as a result, the overall plant cost slightly increases with increase in relative reservoir size.

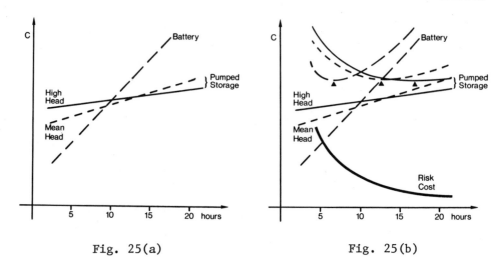

Fig. 25(a)   Fig. 25(b)

Instead, in mean-head plants, the reservoir cost represents a significant share of the overall plant cost; in effect, the reservoirs are generally larger (often there is the need to create reservoirs of the concrete-coated type), while the water supply plants cost definitely less than the high-head plants. As a consequence, the overall cost of a mean-head plant grows significantly with a growing number of hours, i.e. a growing reservoir volume.

In case of batteries, the power is clearly negligible with respect to the energy cost, because what affects the overall cost is the number of batteries, which rises in direct proportion to the number of operating hours required.

Obviously, the calculation of the battery cost took into account the economic advantages that batteries have in transmission and distribution systems, because they can be installed very close to the loads.

To assess the competitiveness between high and mean-head pumped storage plants and batteries, we make an economic calculation.

In fig. 25(b), if we sum up the risk cost (curve in the lower part of the figure) and the mean-head plant cost, we obtain a curve which shows that the optimum value of h is approximately 12 hours. These data correspond rather well to an electric system of the ENEL type, which was planned assuming a storage unit incidence of about 11-12% and requires reservoirs with a capacity sufficient to ensure about 12 hours of production with mean-head plants.

Naturally, if we repeat the economic calculations to define the most advantageous number of hours in case of high-head plants or batteries, we find different values of h: for high-head, the optimum value of h is 16 hours, while for the battery, it is 5-6 hours.

For these values of h, mean-head plants and the battery are competitive because they have really the same cost.

Of course, there are not only high-head plants, mean-head plants, and batteries. In fact, a mix of the three systems is generally required. To have an optimum mix, we will take a value of h greater than h_o in case of high-head plants and a value of h smaller than h_o in the case of batteries.

Also for this reason, it is important that our plants have large reservoirs, because if tomorrow we want to add batteries to such plants, we might do it at relatively modest costs.

Services Performed by a Pumped Storage Plant

So far we dealt with only one of the services that a storage plant can perform, i.e. the power service, which consists in making available the installed capacity whenever it is necessary to meet the demand.

To see whether the storage plant is competitive with turbogas, it is necessary to examine also the other services. To this purpose, we recall that the services that a storage plant can perform within the framework of an electric power production system are not only the static services, i.e. those dependent on the static characteristics of the system (on steady-state operations), i.e. on P, π, and h, but also the dynamic services. The static services include the power service, already mentioned, which serves to ensure the power, and the energy transfer service, which transfers energy from hours when thermal production takes place at lower marginal cost to those where it would take place at higher cost and which results in an operating cost saving. The dynamic services are: The ramp services, consisting in meeting rapid load variations which cannot be covered with conventional thermal power plants (e.g., when in the morning the load rises very rapidly, an entirely thermal system may not follow the rising load). Then, it is necessary to intervene with units which can be rapidly brought to maximum capacity (the pumped storage plant of Vianden prevalently carries out a service of this type). The spinning reserve service consists of meeting unexpected generation disconnections, reducing the consequent outage. And, the frequency control service, which is similar to the previous one and includes, in part, the spinning reserve service.

UNITED STATES NUCLEAR ENERGY AND NON-PROLIFERATION POLICY

Daniel P. Serwer

Scientific and Technological Affairs
United States Embassy
Rome, Italy

I believe that U.S. nuclear energy and non-proliferation policy is not well understood, and I hope to be able to clarify where
the U.S. Government stands on a number of very complex issues.   I
shall speak first about the role of nuclear energy within the context of overall energy policy, then about the development of nuclear energy, and finally about nuclear non-proliferation.   A
general discussion will follow this talk.

Let me start by emphasizing that the Carter Administration
views nuclear energy as a part of the overall energy picture.   No
one, today, can afford to project a completely or even predominant-
ly nuclear future, as was once popular.   In the United States, our
first priority in the near term is energy conservation.   We must
make the transition from an era of cheap oil and gas to an era,
still largely in the future, of expensive oil and gas.   By the
year 2000, we can expect to be using at most not more than 70 percent of the total energy resources that we would have used in the
absence of price increases, and we may well be using much less.

We shall, nevertheless, be using a good deal more energy in
the year 2000 than we are using today, and many projections indicate that an increasing proportion of this energy will be in the
form of electricity.   Nuclear energy is one of our options for
meeting this increased electrical demand.   Our other option is
coal.   Oil and gas are not considered viable fuels for newly
constructed electrical capacity in the United States, except for
"peaking" purposes.   Solar energy, although increasingly seen as
an important source of heat for building and industry, will not
make a major contribution to meeting electrical demand before the
year 2000.

The United States does, however, have a choice between coal and nuclear power, and in this respect we clearly face a different situation from many of our Allies, including Italy. Our domestic coal reserves are enormous; our capacity to utilize them is limited largely by infrastructural requirements, especially railroads, and by environmental considerations.

I am personally convinced that the total health and environmental impact of producing a given amount of electricity with nuclear power is substantially lower in the United States, than the impact of producing that same electricity with coal. In fact, until coming to Rome, I worked at Brookhaven National Laboratory with a research group devoted in large part to quantifying health and environmental impacts of various fuel cycles, and we repeatedly came to the conclusion that coal could be expected to do more harm than nuclear power, even when the best future available pollution control technology was used. I believe that this conclusion is generally accepted in knowledgeable circles within the United States.

Why, you may then ask, are orders for nuclear power plants so slow, and coal much more popular? This question goes to the heart of the matter: the federal government does not directly decide how many power plants of each type will be built in the United States. Unlike the Italian Parliament, the American Congress has never decided on a specific number of nuclear power plants. The government clearly contributes to the atmosphere in which such decisions are made, but it is the publicly regulated electrical utilities--privately or publicly owned--that actually order power plants.

Perhaps the strongest reason for ordering coal rather than nuclear power in recent years has been uncertainty: uncertainty about the regulations that will have to be met in building a nuclear power plant, and about public acceptance of nuclear power in general and of particular sites. These uncertainties have meant delay, and delay in a one-billion-dollar project means added costs. The Carter Administration, following closely on policies already developed under President Ford, has attempted to reduce these uncertainties in two ways: first, by streamlining the licensing procedures; second, by meeting public concerns. I shall only mention briefly the effort to streamline licensing; so far, it has failed to attract significant Congressional support.

The effort to meet public concerns has, I believe, been more successful. Nuclear power is very carefully regulates in the United States, and the public has many opportunities to be involved directly in the procedures of the Nuclear Regulatory Commission. The Commission maintains more than 130 rooms throughout the United States where its documents can be read by the general public.

"Abnormal occurrences" at nuclear power plants that may affect public health and safety are routinely announced to the public. The
public can intervene in many stages of the licensing process, and
citizens can request NRC documents of many types with no justifi-
cation required.

Of course, public participation often leads to uncertainty
and delay, but within limits we are convinced that the price is
a reasonable one. If there is something wrong with the present
provisions for public participation, it lies largely in our failure to create similar procedures for coal-fired power plants,
which are made more competitive by the relative simplicity of
their licensing procedure. In any case, public acceptance of
nuclear power in the United States is much more extensive than is
generally believed in Europe: many referenda have been held at
the State level, and nuclear power has never lost.

One of the intense public concerns in recent years has been
the issue of nuclear proliferation. These concerns arise, in part,
from an exaggerated picture of the role of nuclear power. It was
only a few years ago that nuclear proponents were talking of many
hundreds of power plants and dozens of reprocessing facilities
spread throughout the world. The "plutonium economy," in this
view, was to be a relatively free-trade economy, with each country
doing as it liked under the fairly loose reign of the IAEA Safeguards. The Carter Administration has confirmed the Ford Administration's rejection of this plutonium free-trade economy. It has
done so not in order to hurt nuclear power, but rather to make
nuclear power acceptable. And it has done so not for commercial
advantage, but rather at substantial cost to the American economy.

Let us turn back to April 1977, when President Carter announced
his non-proliferation policy, and review the points he made then.

First, the President emphasized at that time his commitment
to light water reactors without reprocessing. He reiterated this
commitment at the Bonn summit in July 1978: "the further development of nuclear energy is indispensable, and the slippage in the
execution of nuclear energy is indispensable, and the slippage in
the execution of nuclear power programs must be reversed." Clearly,
the Carter Administration is not against nuclear power.

Second, the President deferred indefinitely commercial reprocessing of plutonium in the United States and urged other countries
to do likewise. The cost of this measure to American industry was
substantial, as symbolized in the still unused Barnwell reprocessing plant. This deferral of reprocessing still stands, and I
expect it to continue for two reasons: our experience with commercial reprocessing, from both an economic and an environmental
point of view, has been bad; and reprocessing offers only the

slightest advantages in terms of reducing fuel demand in the next
25 years. If reprocessing and plutonium utilization in light
water reactors are used when they are clearly justifiable on an
economic basis, we believe that they will be very little used for
several decades.

Third, the President reshaped the U.S. program for breeder
reactors by cancelling the planned plutonium liquid metal fast
breeder at Clinch River and increasing funding for more proliferation-resistant technologies. Here one must not be trapped into
misreading U.S. policy. While we are against reprocessing for
recycle of plutonium into light water reactors, we have not yet
reached a conclusion on the issue of the use of plutonium in
breeders. As a result, we maintain by far the largest breeder
reactor development program in the world ($370 million in fiscal
year 1980), and we have every intention of remaining in the fore-
front of this technology. Here the charge that we are seeking
commercial advantage might seem viable, although the American
nuclear industry has bitterly contested the cancellation of the
Clinch River reactor. Nevertheless, U.S. industry may some day
benefit from sales of a more proliferation-resistant breeder technology that the technology under development elsewhere. If this
comes about, however, it will be because there is a genuine demand for such technology, a prospect we can hardly be banking on
given the present response to U.S. non-proliferation policy.

We are sure that less non-proliferation-resistant technology
will prove more popular if the non-proliferation regime is strength-
ened. The President also announced in April 1977 a series of
measures to increase non-proliferation incentives: limited American spent fuel storage for fuel from abroad; nuclear fuel assur-
ances; application of full-scope safeguards as a condition of
U.S. exports; U.S. veto rights over reprocessing of U.S.-supplied
fuel or fuel irradicated in a U.S.-supplied reactor; and restraints
on U.S. export of sensitive technology. None of these measures
significantly improved the commercial position of American companies, and a number of them are thought to have hurt the position
of American industry as well as the nuclear export market as a
whole. The fact of the matter is that these measures had a security goal, not a commercial one: we want to slow the proliferation
of nuclear weapons, and we are willing to pay a substantial price
to do so.

What of the future of American non-proliferation policy:
Where is it headed? The answer to this question depends heavily
on the outcome of INFCE, the International Nuclear Fuel Cycle
Evaluation. INFCE is a joint effort of 50 countries to re-examine
the role of nuclear power. We believe that five basic norms for
a stable non-proliferation regime are emerging from INFCE, norms
that Joe Nye has already spoken about at the Uranium Institute

as well as during his visit to Italy. First is the application of
full-scope safeguards, that is the application of safeguards to
all facilities in a given country as a condition of nuclear exports
by one of the nuclear suppliers. Second, avoidance of sensitive
facilities like reprocessing plants and enrichment plants where
they are not economically justifiable. Third, use of more proliferation-resistant technology. Fourth, joint multi-national control
of sensitive facilities. And fifth, stronger assurance that
needed fuel, enrichment capacity and spent fuel storage will be
available.

These norms certainly forebode a future for nuclear power
quite different from the one its strongest advocates once projected.
Plutonium, if it flows at all, will flow in very restricted and
well-watched channels. Light water reactors will not be built as
rapidly as we once imagined: in the European community, projections
of electricity produced in 1985 by nuclear reactors declined by
40 percent between 1974 and 1978, and even the 1978 projections
are likely to be far too high. We in the U.S. government very
much want such delays to be reversed, but it is no wonder that
when we look at the hard data we feel less urgency about reprocessing and the commercialization of breeder reactors. As George
Rathjens, an M.I.T. professor who now runs the INFCE staff of the
State Department, said when he was in Italy, "a technology can be
developed too early, and the costs of doing so are very substantial." Light water reactors are clearly useful today on an economically justifiable, commercial basis; those who believe the same
will be true in the near term of reprocessing and breeders are
entitled to their view, but I think they stand to lose a bundle.
They should not blame us if they do.

This, in summary, is how the Carter Administration views
nuclear power and non-proliferation for the next two decades:
slower than expected, but substantial, growth in the use of light
water reactors, very limited reprocessing, continued development
of plutonium and other breeders. At the same time we expect much
tighter bilateral and multilateral controls over the entire fuel
cycle, very little commercial reprocessing for thermal recycle
purposes, and a considerably delayed decision on commercialization
of breeder reactors. This is a future different from the one prevalent in the days when enrichment and reprocessing technologies
were being sold to countries before reactors were built, but it is
a future in which the spread of nuclear weapons is likely to be a
good deal slower.

UNITED STATES ENERGY POLICY

Daniel P. Serwer

Scientific and Technological Affairs
United States Embassy
Rome, Italy

There is no greater challenge to the Western world today
than the energy crisis. And during the past year what has seemed
to many to be a difficult but distant problem has come perilously
close to upsetting the world's political, economic and military
balances. I need only mention in passing the revolution in Iran,
the importance to the West of Saudi Arabia, the fall of the dollar
and the risk in gold prices, the current conflict over the Camp
David peace agreements, and the occasional strains in the U.S.-
European relationship over the oil market for us all to realize
the gravity of our energy situation and its repercussions throughout the world.

All of these events are symptoms of a single disease: the
Western world is importing too much oil. The United States bears
a large portion of the responsibility. We are now importing 7.7
million barrels per day (MMBD) or 15 percent of total non-Communist
world consumption. Even with recent increases, American energy
prices are still low by European standards. We are widely, and
with some justice, regarded as profligate users of energy. We
consume nearly twice as much energy per capita as countries like
Germany, which has approximately the same standard of living. We
consume more than three times as much energy per capita as Italy.

These are, however, international comparisons, which count
for little in domestic American politics. It is difficult for
Europeans to imagine the extent to which ordinary Americans are
insulated from international affairs and foreign policy. Until
recently, surveys consistently showed 50 percent of Americans
unaware of the fact that the U.S. imports oil. To the man in the
street, the energy crisis has meant only long lines for gasoline

and sharply higher prices. Even the declining dollar and OPEC
wealth are hardly noticeable to the average American consumer, who
wants fuel at reasonable prices.

True as all this is, it is a mistake to conclude that the
situation is hopeless. Quite to the contrary, the United States
in recent years has moved decisively in the direction of limiting
oil imports. The Carter Administration knows only too well how
impossible it is to fulfill the American consumer's dream. The
era of abundant and cheap oil supplies is gone forever, in the
United States as well as elsewhere. It is hardly surprising that
the President who has brought this message home to Americans finds
himself criticized sharply. The question to be faced in the United
States is no longer whether energy prices will rise or whether oil
imports will be limited, but rather how and when we shall make the
transition to a new, more energy-conservative regime.

On these issues, the President has spoken decisively for the
near-term. While it is generally known that the Congress turned
down the President's effort in 1977 and 1978 to raise domestic
American oil prices to world levels using a tax, it is less generally recognized that since then the President has decided to
exercise his executive powers to achieve the same goal by decontrolling domestic oil prices. This decision does not require Congressional approval.

As of June 1 of this year, newly discovered American crude
oil, as well as oil produced with advanced recovery techniques,
is being sold at world prices. Beginning January 1, 1980, presently
controlled crude oil prices will rise gradually to world levels.
Decontrol will be completed by September 30, 1981.

In addition to oil price decontrol, the President has announced
that American oil imports will not be permitted between now and
1985 to rise above 8.5 MMBD, the level reached in 1977. The very
sharp increase in American oil imports that has caused so much
difficulty in the 1970s is at an end. If necessary, the President
will use his executive powers to impose quotas limiting the amount
of oil entering the United States.

This basic constraint on oil imports still leaves us with a
good deal of freedom in the longer term. Indeed, the great challenge posed to the West by the energy crisis is the challenge of
long-term projections and planning. What fuels will we be using
in the year 2000 and beyond? How much of these fuels can be displaced by solar energy and by a greater commitment to energy conservation? How will nuclear power develop?

Let me turn to this longer-term transition, first by taking
a closer look at American energy consumption in an economic context

and then by reviewing the very extensive legislation--both in
effect and proposed--that we believe will make an orderly transition possible.

   While it is true that per capita American energy consumption
is very high, some of this energy consumption is due to the high
productivity of our economy. Americans consume more energy partly
because they produce more goods and services per capita. Despite
our high per capita energy consumption, our energy consumption per
dollar of gross domestic product is actually lower than that of
the United Kingdom, and only 25 percent higher than that of Italy.

   Moreover, we have in the last few years been able to reduce
significantly the amount of fuel we need to produce a dollar of
economic output. The rate of increase of our energy consumption
was only 1.8 percent in 1978, down from 2.5 percent in 1977 and
5.3 percent in 1976, all years of economic growth. In the years
1962-72, when real energy prices in the U.S. declined by 1.4 percent per year, economic growth in the U.S. brought a more or less
proportional increase in energy usage. This is no longer true,
largely because real energy prices from 1973 to 1978 have increased
at a rate of 12.5 percent per year. This year, American energy
prices have already increased 29.8 percent in the first nine
months of 1979. Our oil consumption in the first half of 1979
actually declined by one percent, and our gasoline consumption
has declined by 3.4 percent. For every one percent increase in
GNP, we now have much less than a one percent increase in energy
consumption: an increase of only .5 percent appears to be sustainable, although in fact recent increases have been even lower.

   In many countries of Western Europe, economic growth and
energy use are still more or less proportional. Thus per capita
U.S. energy consumption is likely to remain high. But consumption
per dollar of GNP may well approach the level of Western Europe in
the longer term.

   To summarize, we regard continued economic growth as critical
in the long term. We have no intention of saving energy by reducing growth. The social impact would be primarily on the poor,
and we need growth in order to invest in the capital required for
our longer-term energy transition. However, we believe that growth
in energy consumption can be much less than it has been in the
past, and we plan to use both price and non-price mechanisms to
limit the energy effects of economic growth.

   Looking beyond oil price decontrol, the key to these mechanisms, we find a very complex set of legislation. Some of it has
been approved by the Congress and signed into law by the President.
Other items are still Presidential proposals, and their fate in

the Congress is, of course, uncertain. Let me review this legislation for you.

     After a year-and-a-half of Congressional discussion, the
President signed five bills into law in November 1978:

     1.   The National Energy Conservation Policy Act;

     2.   The Power Plant and Industrial Fuel Use Act, essentially
a "concoal" conversion law;

     3.   The Public Utilities Regulatory Policy Act;

     4.   The Natural Gas Policy Act; and

     5.   The Energy Tax Act.

The only essential proposal the President made in April 1977 that
failed to pass the Congress entirely was a tax on domestically
produced oil to raise its price to the world price, a goal that
he has now implemented by decontrolling oil prices.

     Let me offer an overview of these five acts already in effect
in the U.S.

ENERGY CONSERVATION

     This Act offers standards and subsidies for conservation, with
a strong bias toward grants and loans for conservation to lower-
income families as well as conservation and solar energy in public
buildings.

COAL CONVERSION

     This Act prohibits new oil and gas boilers for electrical or
industrial use and restricts use of oil and gas as a fuel in existing boilers, thus making coal the boiler fuel of choice for the
future. Exemptions in special circumstances can be made.

PUBLIC UTILITY REGULATORY POLICIES

     This Act encourages adoption by gas and electrical utilities
of more energy-conservative rate structures, requires utilities
to buy and sell power to industrial cogenerators, encourages grid
interconnections, and provides loans for the development of small
hydroelectric projects.

NATURAL GAS PRICING

This Act provides for the decontrol of most newly-discovered gas by 1985, but gas from old wells will remain under control.

ENERGY TAXES

This Act offers income-tax credits to individuals:

a) for residential conservation, the credit is $300 or 15% of the first $2000 expended;

b) for residential solar or wind devices, the maximum credit is $2200 for up to $10,000 of expenditure.

The Act also severely taxes "gas guzzling" cars and exempts "gasohol," gasoline mixed with alcohol produced from agricultural residues, from the Federal Excise Tax on gasoline.

The estimated 1985 savings of imported oil as a result of this legislation already in force, is as follows:

Conservation	.7 million barrels/day
Utility rates	.2
Coal conversion	.3
Taxes	.4
Natural gas pricing	1.0
TOTAL	2.5

These are quite substantial savings, amounting as they do to 32 percent of our current oil imports.

These are large savings, but they are far from sufficient. Even adding the 1.5 million barrels per day that we estimate to be the savings by 1990 from oil price decontrol plus further measures to encourage solar energy, we would still be only managing to keep our oil imports at more or less current levels rather than actually reducing them. The President has therefore proposed a series of additional measures, funded by a tax on the additional oil company profits due to the decontrol of oil prices. This tax, if enacted by Congress, will provide at least $146,000 million between 1980 and 1990.

As proposed by the President, this fund would be split in
two big pieces. The smaller piece, about $50,000 million would
be spent by 1990 primarily on further support for energy conservation and solar energy, including a solar energy bank to provide
low-cost loans to home-owners who want to install solar devices
and substantial assistance to people with low incomes as well as
to mass public transportation. While the President's proposals
may be changed by the Congress, I think there is relatively little
controversy about the wisdom of using a substantial portion of the
so-called "windfall" profits tax for energy conservation and solar
energy.

The larger part of the proceeds, about $90,000 million, would
be spent by 1990 on producing synthetic and unconventional fuels.
Here there is more controversy. The Administration has already
accepted Congressional initiatives that would slow down the initial
proposal, in essence making the Administration's goal of 2.5 MMBD
effective for the year 2000 rather than 1990. The essence of the
proposal would remain the President's: a government-owned Energy
Security Corporation would use public funds to offer price guarantees, to create a synthetic fuels industry. These fuels would
include liquids and gases made from coal, oil from oil shale, gas
from agricultural products and waste, and gas from so-called "tight"
sands."

Why undertake such a large, government-sponsored program for
synthetic fuels? And is 2.5 MMBD in the year 2000 worth the expenditure of $90,000 million in public funds, plus much more in
private inventments?

There are, I think, several good reasons for this new program.
First, America's energy problems are largely problems of liquid
fuels. We have already prohibited the use of oil in new industrial
and utility boilers, and the President will soon send proposals to
the Congress for early retirement of existing oil-fired boilers.
In order to make further headway on liquid fuels, we are virtually
forced to use coal, shale, biogas, and other unconventional gases
to substitute for oil in transportation and in heating buildings.

Second, the government must act because private industry,
given current oil prices, cannot justify major unsubsidized investments in synthetic fuels. Oil shale will cost $28-$35 per barrel
and gas and oil from coal will cost $35-$40 per barrel. Current
world prices are $20-$22 per barrel, and controlled American prices
still substantially lower, but we know full well that both American
and world prices will rise. Public intervention is necessary if
we are to move ahead with synthetic fuels as insurance against
these price increases.

And finally, we must, I think, begin to move now rather than waiting. Predictions of oil prices are impossible. We know, however, that, even in the best of all possible worlds, with Iranian production fully restored and no further surprises, that the mid-1980's to mid-1990's will be a period of great strain in the oil market, with supply at current prices falling short of demand. To invest $90,000 million of public funds to produce 2.5 million barrels of oil per day in 1990 or 2000 would not in itself be worthwhile. If, however, one believes that by the 1990's the synthetic fuel plants will be operating on an economic basis, then this massive public investment looks very reasonable indeed.

Let me turn back to the big picture and look at overall American energy supply in the year 2000. One can, I think, offer a projection based on the Carter Administration's energy initiatives. Total American energy use is likely to grow much more slowly than in the past, averaging about one percent per year. Coal use will double, making it by far our greatest single energy source in the year 2000. Gas and oil supplies will barely manage to hold their present levels. Nuclear energy will grow substantially to about 14 percent of total energy supply, both because we shall finish the 130 nuclear plants now under construction (70 are already operating) and because we hope to see orders resume again before the mid-1980's. And solar energy, which includes in our definition wind, hydroelectric power and biomass fuels as well as the more obvious solar technologies, will increase at a somewhat slower but substantial rate.

Last, but by far not least, we hope to cut American oil imports to four MMBD by the year 2000, half the current figure. This compares with projections, in the absence of vigorous measures to promote conservation and alternative energy sources, as much as four times higher.

We cannot afford to let any more time slip by. I am convinced that the United States has begun the long and difficult process of conserving energy and limiting its oil imports. We cannot even hope for energy independence, as once we thought possible. But with our domestic house now at least partly in order, we need to turn to our Allies and friends to form an even stronger coalition of oil consuming countries. As our Secretary of Energy said at the Paris conference of Energy Ministers: "We share a collective responsibility for the orderly and timely development of a stable and secure energy market. I am confident that we have the means and the political will to meet this responsibility.

ENERGY SCENARIOS FOR THE U.S.

Richard Wilson
Harvard University
Cambridge, MA 02138

RECENT U.S. ENERGY HISTORY

In this lecture I will talk about the U.S. only and this is
both a strength and a weakness. It is a strength because the U.S.
maintains good statistical records and a weakness because international issues pay an increasingly important part in the energy
policy of the Western countries.

In order to explain the usefulness of constructing scenarios,
let me first describe the usual way of proceeding before 1970. We
plotted the logarithm of the energy use against time as shown in the
sketch:

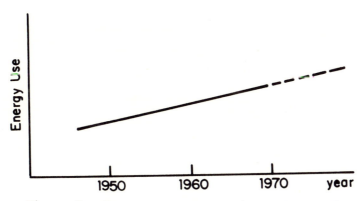

Figure 1.   Energy use versus time before 1970.

and extrapolated into the future.   From 1950 to 1970 such a plot
demonstrated increases as follows:

economy as a whole	3.2%/year
energy	3.4%/year
electricity consumption	9.0%/year

The close link between the economy and energy was in many circles
regarded as fixed.   Two figures illustrate this: one a relation
between GNP of a country and its energy use (Figure 2) and the
next (Figure 3) a <u>differential</u> graph showing how changes in energy
use and changes in GNP have been highly correlated in a short time
scale.

     However, the correlation is not complete, as a plot of the
ratio of energy use to GNP in different countries (Figure 4) shows.
The question is posed: can we <u>reduce</u> the historical ratio for the
U.S. to the European ratio without lowering the quality of life?

     Of the increase in energy use, about 2%/year was due to popu-
lation increase; the switch toward electricity was regarded as a
trend toward more convenience and less environmental pollution.
The population increase has now slowed down and we now regard

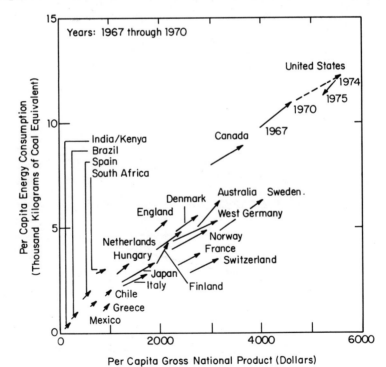

Figure 2.   Per capita energy consumption versus gross national
product for a number of countries.

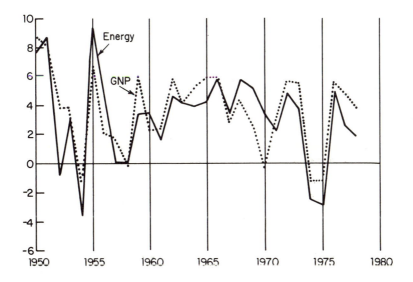

Figure 3.   Changes in Primary Energy and GNP, 1950-1978.

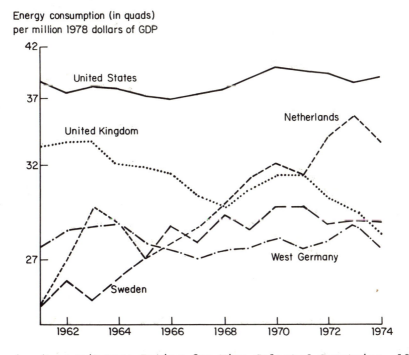

Figure 4.   Energy/Output Ratios for Five Selected Countries, 1961-1974.

any further switch toward electricity, if it occurs, as a change in
the source of fuel.

These projections were already beginning to become erroneous
before 1973 when the dramatic increase in oil prices came. However,
there were people in the U.S. who did have farsighted plans. It is
useful to remember what they were, and why the plans have been rejected or modified.

The trend toward electricity use was taken very seriously.
The price of electricity had, in real terms, been steadily falling,
and electrical home heating was becoming a reality, particularly
for apartment dwellers, where the ease of individual control out-
weighs cost disadvantages.

There was overoptimism about nuclear fission and fusion electric power. One careless remark--"electricity will be too cheap
to meter"--has often been quoted. Some of the technical optimism
was misplaced, but much more serious was optimism about public
acceptance.

The Connecticut Yankee Atomic Power plant was put into service
in 1967 and has been a very successful plant. In 1970 it was producing electricity at 0.55 cents per kilowatt hour including interest and amortization of capital and fuel costs. (It is now less
than 1.5 cents/kwh). This was competitive with the fuel cost alone
of an oil fired generating plant, given 40% efficiency and the cost
of oil at $3.75 a barrel. Residual oil was being bought at $1.75
a barrel on the East Coast of the U.S., but prices were expected
to increase.

It was therefore confidently expected that nuclear power would
replace most oil for electricity generating just as soon as it could
be built. Plans in my area, New England, (the Zinder report), anticipated that by 1990, 80% of all electricity would be generated by
nuclear power; plans were also being made to produce process steam
by nuclear power in large industrial areas--such as Midland, Michigan.

In areas with coal fields, coal was expected to keep competitive with nuclear power--since in these areas the cost of transportation (which doubles the cost of coal in New England and quad-
ruples it in Japan) is small.

It is common now to state categorically that these pre-1970
plans and projections were wrong and ignorant. I think this shows
a lack of understanding of the energy situation.

Society has made the following new demands on a nuclear power
station:

   1.   Increased attention to environmental matters since the
National Environmental Policy Act of 1969, often termed the most
important energy legislation of the century.   No longer can power
stations have simple once-through cooling.   They now must have
cooling towers, or cooling water discharges 2 miles out in the
ocean.

   2.   Radioactivity release in normal operation has been reduced ten-fold.

   3.   Increased public participation makes public hearings
last 5 years instead of 1 day.

   4.   Antinuclear groups, for a variety of reasons, have insti-
tuted deliberate delaying tactics.   For example, a delay of 18
months in the awarding of an operating license to Vermont Yankee
Nuclear Power Plant meant that the plant was idle for this time,
while banks and lawyers continued to be paid.   This probably increased the capital cost 25%.

   5.   Somewhat associated with 4, greater regulatory requirements by the Nuclear Regulatory Commission, some associated with
safety, but some not.

   6.   Increased cost and declining productivity of U.S. construction labor.

   7.   Increases due to the rising costs of materials (inflation)
and rising interest rates which are an expression of the expectation
of further inflation.   Sometimes these are included as costs of
delay, but they are really separate.

   At a very rough guess I would put increased costs at the
following:

Table 1

Environmental	x 1.3
Delays caused by increased public participation and opposition	x 1.3
Increased regulatory requirements	x 1.2
Increased labor costs relative to the rest of the economy	x 1.2
Inflation and expectation of inflation	x 1.8
Overall capital cost increase	x 4.4

In our discussions we should logically try to separate out inflation and expectation of inflation--but this is hard to do and is rarely done consistently.

It is not my purpose here to argue that any one of these causes of price increase are wrong in any absolute sense. I supported improved environmental controls. But they have occurred and to a very large extent they were deliberate decisions made by the American public.

Since 1973 cost increases have been great all over the Western world. I now go to a paper by my colleague, Prof. William Hogan of the Kennedy School of Government at Harvard.[1] Hogan first notes that reactions to increased energy costs will depend on net energy costs delivered not on primary energy costs. In the U.S. there has historically been a marked difference betweeen the two which is illustrated in Figure 5. As the primary cost rises and the delivery cost stays the same, the difference has been going down. (These prices are corrected for inflation to 1978 dollars.)

The important point is that by 1979, the increase in delivered cost was only 20%; this is not enough to stimulate much reduction in energy demand.

However, we can see this broken down by sector in the next table. Here I show delivered energy prices for each of 3 primary sources--solid (coal), oil, and gas--and for electricity in industry, transportation and residential for 3 years. In the accompanying Table 3 we see the energy actually used.

We see at once that energy prices, as delivered to industry, rose rapidly after 1973 and that industry responded rapidly by reducing demand. On the other hand the delivered prices in the domestic sector had a smaller fractional risk, and the demand barely changed. The switch away from gas, toward oil, in industrial use was in response to a government directive. Noteworthy is a doubling of domestic electricity use in this decade. Yet it is in this sector that we hear cries for reducing energy demand and electricity demand in particular.

Similar tables for the U.K. are shown in Tables 4 and 5; note the increased use of gas in the U.K. due to inexpensive North Sea (subsidized) gas.

Now I turn to scenarios for the future, particularly those suggested by the CONAES study. In this talk I will not refer to this study directly, but to a published summary.[2] The change which is forced by the inability of domestic oil supply to meet demand, with the trend to foreign oil and a high price, suggests a re-examination of the whole energy system and its impact on society.

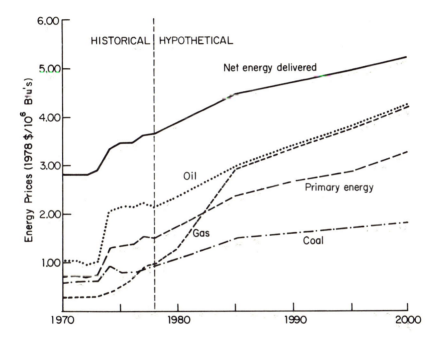

Figure 5.   Illustration of Relation between
Primary-Energy Prices and Net-Energy Delivered Prices

TABLE 2

Prices by Fuel and Sector, United States*

Industry

Year	Solid Fuels	Oil	Gas	Electricity
1968	0.399	0.619	0.478	4.840
1969	0.394	0.585	0.462	4.671
1970	0.423	0.644	0.464	4.623
1971	0.467	0.858	0.475	4.797
1972	0.481	0.920	0.490	4.911
1973	0.488	1.099	0.523	4.990
1974	0.705	2.351	0.677	6.015
1975	0.814	2.136	0.930	6.805
1976	0.795	2.025	1.199	6.954
1977	0.821	2.129	1.441	7.438
1978	0.881	1.945	1.489	7.541

Transportation

Year	Solid Fuels	Oil	Gas	Electricity
1968	0.000	5.155	0.000	0.000
1969	0.000	5.087	0.000	0.000
1970	0.000	4.940	0.000	0.000
1971	0.000	4.825	0.000	0.000
1972	0.000	4.630	0.000	0.000
1973	0.000	4.709	0.000	0.000
1974	0.000	5.785	0.000	0.000
1975	0.000	5.714	0.000	0.000
1976	0.000	5.665	0.000	0.000
1977	0.000	5.744	0.000	0.000
1978	0.000	5.817	0.000	0.000

Residential/Commercial

Year	Solid Fuels	Oil	Gas	Electricity
1968	3.302	2.284	1.528	11.312
1969	3.240	2.239	1.486	10.694
1970	3.079	2.206	1.480	10.208
1971	2.999	2.256	1.528	10.246
1972	2.967	2.541	1.572	10.413
1973	2.880	3.811	1.560	10.361
1974	3.276	3.723	1.633	11.018
1975	3.108	3.358	1.844	11.408
1976	3.096	3.389	2.097	11.586
1977	3.156	3.696	2.440	12.160
1978	3.216	3.686	2.572	12.263

*All prices are $1978/MMBTU

TABLE 3

Consumption by Sector and Fuel, United States*

Industry

Energy Demand (Delivered) in Quads

Year	Solid Fuels	Oil	Gas	Electricity	Total
1968	3.92	5.16	9.66	2.58	21.32
1969	3.91	5.24	10.33	2.76	22.24
1970	3.78	5.59	10.69	2.90	22.96
1971	3.45	5.69	11.02	3.01	23.16
1972	3.83	6.00	11.14	3.19	24.17
1973	4.14	6.21	10.78	3.35	24.48
1974	3.82	5.67	10.90	3.31	23.69
1975	3.44	5.53	9.34	3.20	21.50
1976	3.49	5.68	9.20	3.47	21.83
1977	3.32	6.00	8.91	3.63	21.86
1978	3.14	6.26	8.37	3.74	21.51

Transportation

Energy Demand (Delivered) in Quads

Year	Solid Fuels	Oil	Gas	Electricity	Total
1968	0.02	13.51	0.00	0.02	13.55
1969	0.02	14.06	0.00	0.02	14.09
1970	0.02	14.53	0.00	0.01	14.56
1971	0.01	15.20	0.00	0.01	15.23
1972	0.00	16.16	0.00	0.01	16.18
1973	0.00	16.87	0.00	0.01	16.88
1974	0.00	16.69	0.00	0.01	16.71
1975	0.00	17.21	0.00	0.01	17.22
1976	0.00	17.78	0.00	0.01	17.79
1977	0.00	18.29	0.00	0.00	18.29
1978	0.00	19.07	0.00	0.00	19.07

*Continued on following page.

TABLE 3 (continued)

Residential/Commercial

Year	Solid Fuels	Oil	Gas	Electricity	Total
1968	0.50	5.56	6.54	2.13	14.74
1969	0.42	6.23	7.00	2.34	15.99
1970	0.40	6.54	7.63	2.48	17.05
1971	0.38	6.55	7.72	2.65	17.29
1972	0.34	6.98	7.93	2.94	18.19
1973	0.31	6.42	8.32	3.16	18.21
1974	0.28	5.99	7.62	3.37	17.26
1975	0.27	5.22	7.72	3.60	16.81
1976	0.27	6.75	8.27	3.76	19.05
1977	0.27	7.15	8.01	3.98	19.40
1978	0.30	7.43	8.77	4.19	20.69

Total Energy Demand (Delivered) in Quads, by Fuel Type

Year	Solid Fuels	Oil	Gas	Electricity	Total
1968	4.44	24.23	16.21	4.73	49.61
1969	4.35	25.53	17.32	5.12	52.32
1970	4.20	26.66	18.31	5.40	54.57
1971	3.84	27.44	18.73	5.67	55.68
1972	4.18	29.14	19.08	6.15	58.54
1973	4.45	29.50	19.10	6.53	59.57
1974	4.10	28.35	18.52	6.69	57.65
1975	3.71	27.96	17.05	6.81	55.53
1976	3.75	30.21	17.47	7.24	58.67
1977	3.59	31.43	16.92	7.61	59.55
1978	3.44	32.76	17.14	7.93	61.27

TABLE 4

Prices by Fuel and Sector, United Kingdom*

Industry

Year	Solid Fuels	Oil	Gas	Electricity
1968	1.211	1.461	3.999	11.314
1969	1.106	1.385	3.323	10.719
1970	1.050	1.079	2.393	10.145
1971	1.152	1.567	1.588	10.261
1972	1.273	1.528	1.330	9.692
1973	1.268	1.313	1.652	9.049
1974	1.178	1.633	1.289	11.042
1975	1.846	4.053	1.368	12.353
1976	2.148	3.870	1.539	13.315

Transportation

Year	Solid Fuels	Oil	Gas	Electricity
1968	0.000	10.837	0.000	0.000
1969	0.000	11.647	0.000	0.000
1970	0.000	11.193	0.000	0.000
1971	0.000	10.674	0.000	0.000
1972	0.000	10.027	0.000	0.000
1973	0.000	9.539	0.000	0.000
1974	0.000	10.178	0.000	0.000
1975	0.000	18.307	0.000	0.000
1976	0.000	16.068	0.000	0.000

Residential/Commercial

Year	Solid Fuels	Oil	Gas	Electricity
1968	2.918	3.073	6.471	15.370
1969	2.778	2.866	5.992	14.080
1970	2.813	2.657	5.601	12.999
1971	2.944	2.949	4.927	12.891
1972	3.180	3.131	5.106	12.815
1973	3.193	3.624	5.458	11.947
1974	2.796	3.384	5.089	12.377
1975	2.706	5.424	4.596	12.536
1976	3.015	5.916	4.599	14.687

*All prices are $1978/MMBTU

TABLE 5

Consumption by Sector and Fuel, United Kingdom*

Industry

Energy Demand (Delivered) in Quads

Year	Solid Fuels	Oil	Gas	Electricity	Total
1968	1.16	1.57	0.16	0.36	3.26
1969	1.15	1.71	0.20	0.39	3.43
1970	1.08	1.80	0.26	0.40	3.53
1971	0.94	1.84	0.36	0.40	3.54
1972	0.80	1.92	0.51	0.40	3.64
1973	0.79	2.00	0.59	0.44	3.82
1974	0.71	1.80	0.61	0.41	3.54
1975	0.66	1.52	0.60	0.41	3.20
1976	0.65	1.57	0.64	0.43	3.29

Transportation

Energy Demand (Delivered) in Quads

Year	Solid Fuels	Oil	Gas	Electricity	Total
1968	0.00	1.03	0.00	0.00	1.03
1969	0.00	1.07	0.00	0.00	1.07
1970	0.00	1.12	0.00	0.00	1.12
1971	0.00	1.16	0.00	0.00	1.16
1972	0.00	1.31	0.00	0.00	1.21
1973	0.00	1.29	0.00	0.00	1.29
1974	0.00	1.24	0.00	0.00	1.24
1975	0.00	1.23	0.00	0.00	1.23
1976	0.00	1.27	0.00	0.00	1.27

---

*Continued on following page.

TABLE 5 (continued)

Residential/Commercial

Energy Demand (Delivered) in Quads

Year	Solid Fuels	Oil	Gas	Electricity	Total
1968	1.03	0.41	0.35	0.32	2.11
1969	0.97	0.46	0.39	0.35	2.17
1970	0.89	0.49	0.43	0.37	2.18
1971	0.74	0.50	0.48	0.38	2.10
1972	0.65	0.53	0.55	0.41	2.14
1973	0.63	0.55	0.59	0.43	2.21
1974	0.60	0.48	0.67	0.43	2.18
1975	0.51	0.48	0.73	0.42	2.14
1976	0.47	0.49	0.78	0.42	2.15

Total Energy Demand (Delivered) in Quads, by Fuel Type

Year	Solid Fuels	Oil	Gas	Electricity	Total
1968	2.19	3.02	0.51	0.69	6.40
1969	2.12	3.23	0.59	0.73	6.67
1970	1.96	3.41	0.69	0.76	6.83
1971	1.68	3.50	0.84	0.79	6.81
1972	1.45	3.66	1.06	0.81	6.99
1973	1.43	3.84	1.18	0.87	7.31
1974	1.31	3.53	1.28	0.84	6.96
1975	1.17	3.23	1.33	0.84	6.56
1976	1.12	3.33	1.41	0.85	6.71

On the one hand, as we planned before 1970, we could again choose a rapid but smooth transition to nuclear energy or coal for all uses of energy in stationary devices--electricity genera- tion, industrial heat and home heat--leaving oil for transportation. On the other hand, we could try to reduce demand and develop alter- native sources of supply so that both nuclear power, with the pro- blems it raises, and coal, with its health and environmental effects can be avoided. The CONAES report studies some of these scenarios as listed in Table 6.

Table 6

Scenario	GNP growth rate (aver. ann.%)	Energy price ratio 2010/1975(aver.)	Energy conservation policy
I_2	2		Very aggressive, deliber- ately arrived at reduced demand requiring some lifestyle changes
I_3	3	4	
II_2	2		Aggressive; aimed at maximum efficiency plus minor lifestyle changes
II_3	3	4	
III_2	2		Slowly incorporates more measures to increase efficiency
III_3	3	2	
IV_2	2		Present policies unchanged
IV_3	3	1	

Firstly, we have to decide what we would like our equilibrium energy position to be--choosing from among realistic alternative strategies. Secondly, we have to decide how to get there from here. In discussing this, CONAES make several assumptions. Among these is that it is politically impossible to return to the pre-1970 forecasts of costs and expansion capability of nuclear power and coal. The delays and additions caused by the need to protect the environment, and even more important to satisfy the public that the environment is protected, have increased costs much higher than inflation.

Thirdly, we have to decide that a completely free market in energy is politically impossible now, even if it existed or were possible at one time.

Fourthly, we have to decide that energy costs in "constant" dollars will double in 20 years; this is very likely for oil and gas which we are running out of, but it is also assumed for the capital construction of a nuclear or coal plant. The basis for this assumption is the experience of the last 8 years that I outlined earlier, but it can be questioned; do we now have enough safety or enough stack gas purification?

Fifthly, CONAES assume the principal political method for influencing demand will be price adjustment. If the price goes up, demand will go down. Price can be raised higher than cost, either by taxes or by profits.

I show in Figure 6 from the CONAES study how to achieve the various energy scenarios.

For scenarios II_a, II_b it is assumed that energy prices rise a factor of 4 in attempts to reduce the demand. For scenarios I_a I_b, it is assumed in addition that aggressive laws and regulations are used to force lower energy demand. We already saw that energy prices in the U.S. are only about 70% of those in Europe, so this corresponds to a price risk to 2 1/2 times the European levels.

I note from Figure 2 it is possible to have self consistent energy scenarios with GNP growth rates of 2% or 3% per year.

I would, in line with my earlier discussion, include a scenario V_3: a return to the 1970 plan with more nuclear power, less constraints and therefore lower price (0.8). This is the 1970 projection with delays. I would do this to emphasize what we have already decided to give up.

The point of this part of the CONAES study is that the choice between the desired outcomes is a political one and the means that are taken now to influence the outcome are political decisions.

The next table (7) shows the breakdown by energy sources for several scenarios. It is worth commenting on the ways in which the explicit, and hidden, assumptions enter into the different results.

In none of the scenarios is there an abolition of oil imports. Yet other countries regard this as the biggest single action which the U.S. can take. Continued oil imports keep up the world price and make oil unavailable for less developed countries. International policies demand that this attention to oil imports be a prime policy statement for the U.S. government. In Tokyo, the French President, Mr. Giscard D'Estaing and the Prime Minister of France urged this very strongly on President Carter. Although I believe this is a very important point, the main new feature of CONAES--a serious discussion of our energy scenarios--is unaltered by such

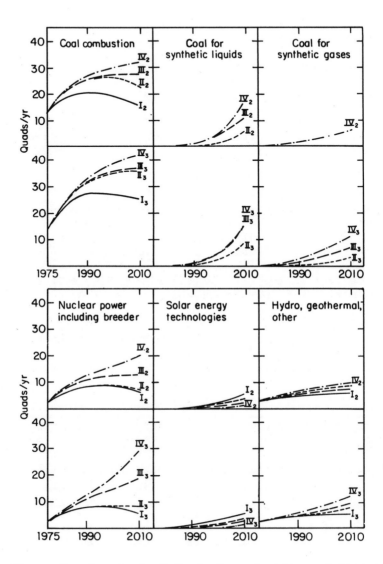

Figure 6.   Trends in U.S. fuel production and imports,
            adopted for the CONAES scenarios.

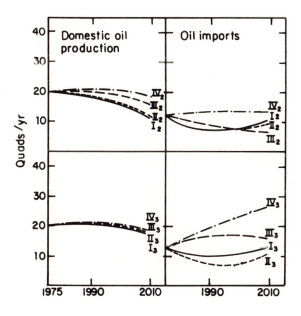

Figure 6 (continued)

TABLE 7

CONAES Scenario I (fuel mix in quads/year)

Scenario	1975	1990	2010
I_2: 2% GNP growth			
Oil			
domestic	20	18	11
imported	13	5	12
shale	0	0	0
Gas			
domestic	19	13	8
imported	1	0	0
Coal			
combustion	13	20	15
conversion to synthetic liquid	0	0	0
conversion to synthetic gas	0	0	0
Nuclear	2	8	6
Solar	0	1	6
Other (hydro, geothermal, etc.)			
Total	71	70	64
Liquid Fuels[a]	33	23	23
Gaseous Fuels[a]	20	13	8
Electricity[b]	20	25	17
II_2: 2% GNP growth			
Oil			
domestic	20	18	11
imported	13	5	10
shale	0	0	1
Gas			
domestic	19	13	10
imported	1	0	3
Coal			
combustion	13	25	22
conversion to synthetic liquid	0	0	6
conversion to synthetic gas	0	0	0
Nuclear	2	8	7
Solar	0	1	4
Other (hydro, geothermal, etc.)			
Total	71	76	83
Liquid Fuels[a]	33	23	26
Gaseous Fuels[a]	20	13	13
Electricity[b]	20	31	29

[a] Includes losses in production and distribution, but not conversion for synthetic fuels derived from coal.

[b] Includes conversion losses.

TABLE 7 (continued)

Scenario	1975	1990	2010
III_3: 3% GNP growth			
Oil			
domestic	20	21	18
imported	13	16	14
shale	0	0	2
Gas			
domestic	19	14	14
imported	1	1	1
Coal			
combustion	13	29	34
conversion to synthetic liquid	0	2	19
conversion to synthetic gas	0	1	7
Nuclear	2	11	18
Solar	3	5	10
Other (hydro, geothermal, etc.)			
Total	71	101	140
Liquid Fuels[a]	33	38	47
Gaseous Fuels[a]	20	16	20
Electricity[b]	20	38	48
IV_3: 3% GNP growth			
Oil			
domestic	20	21	18
imported	13	20	27
shale	0	0	3
Gas			
domestic	19	18	16
imported	1	0	6
Coal			
combustion	13	31	42
conversion to synthetic liquid	0	1	19
conversion to synthetic gas	0	3	12
Nuclear	2	12	30
Solar	0	0	2
Other (hydro, geothermal, etc.)			
Total	71	113	188
Liquid Fuels[a]	33	42	61
Gaseous Fuels[a]	20	20	30
Electricity[b]	20	45	71

[a]Includes losses in production and distribution, but not conversion for synthetic fuels derived from coal.

[b]Includes conversion losses.

an adjustment.

A hidden assumption of scenarios I_2 and I_3 is that nuclear power is undesirable. Indeed, it seems that a major motivation of those urging these scenarios is a desire to avoid completely the use of nuclear electric power. We see in I_2, therefore, a reduction in nuclear electric power by the year 2010. We could, if we choose, modify scenarios I_2 or I_3, increase nuclear electric power and reduce oil consumption somewhat more. Then we can completely avoid oil imports.

Economically, the assumption that the costs of nuclear power will rise faster than inflation--and as fast as oil prices--tends to lead us to this dependence on oil. However, if nuclear power costs rise no faster than inflation, economic considerations will lead to more nuclear power.

I have talked privately to many of the persons who have been vocal and effective in public discussions of alternatives to nuclear power. They all privately agree that their proposals imply this increase of fuel price by a factor of 4, although for tactical reasons they do not always say it in public.

I want, therefore, to disucss some aspects of this factor of 4 price increase. This will not all be a cost increase, and therefore for purposes of calculating the change in gross national product, all of this factor is not involved.

The most important conclusion of the CONAES study is that moderation of energy demand is of the highest importance. I list in order the following conclusions in this regard:

1. Very substantial moderation in energy demand growth is realizable over a 20-30 year period as a result mainly of technical efficiency improvements in the use of energy without adverse effects on economic growth or employment.

2. Even at present energy prices, there are many opportunities for investments in energy-use efficiency that cost less per unit of energy saved than the investment cost of increasing energy supply by the same amount.

3. The potential for cost-effective energy conserving investment increases with increasing energy prices. Conservation could therefore be stimulated by energy taxes.

4. If the economically optimal response to energy price changes is to be realized in practice, especially by individual consumers and households, it will probably have to be encouraged by mandatory performance standards for durable goods, including housing.

5. A wide range of future growth rates in energy consumption
is technically feasible and compatible with continued high economic
growth and little change in consumption patterns in the nonenergy
sectors. Which path should be followed is primarily a question
for political and social choice. Political difficulties tend to
increase at both the high and low end of the energy growth possibilities.

6. Sustained and predictable price and regulatory trends are
necessary to the realization of any of the projected growth paths
without social and economic disruption.

7. Continued research on the economic and distributional consequences of various tax, pricing, and regulatory practices and
policies is necessary to improve the basis for energy conservation
policy.

It is interesting to discuss what the effect would be on GNP
of a _reduction_ of energy use. The pre-1970 conventional wisdom
would imply a proportionate reduction in GNP. One important change
in perception is that the historical energy/GNP ratio might be
halved by technical efficiency improvements. This is discussed
in detail in a paper by Jorgenson and Hudson.[3]

The point is that the increase in _total_ costs is in not just the
direct use of energy, but also the indirect use of energy. Jorgenson uses an input/output analysis to trace energy and capital flows
through the economy. In Table 8 I show the assumptions of the 4
policies studied and in Figure 7 I show the way in which the calculation proceeded.

Finally, in the Table 9 I show some results of their calculations. They list alternative procedures by which they suggest that
energy consumption be reduced. The base case is the 1979 set of
laws and regulations, and the output is listed as percentage changes
resulting from these laws. Let us look at Jorgenson's proposal 4.

This is designed to bring the U.S. energy consumption down to
70 Q (10^{16} Btu per year), which is about the level of the CONAES
scenario I. They agree with CONAES that energy prices have to
increase a factor of 3 1/2. Relative to cost of capital (set as
1.0) the labor costs drop to 0.9 of the base case value; this is
a reflection of an assumption that less skilled (and more) labor
is used for energy conserving devices than for energy production
devices.

Energy costs account for about 5% of the U.S. gross national
product. Scenario I of CONAES suggests an increase of a factor
of 4 in energy costs. Jorgenson and Hudson consider the effect
of an increase of a factor of 3 1/2 (not 4). Naively one might

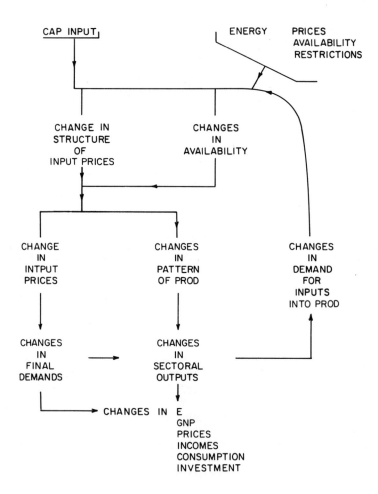

Figure 7.   Flow chart of the Jorgenson Calculation

TABLE 8

Jorgenson Calculation

Base Case	1979 (June)
Policy 1	Taxes on domestic oil to world energy prices Energy conservation taxes and subsidies Oil and gas conversion to coal
Policy 2	+ tax on oil to $4.50/BBL in 1985 $7.00/BBL in 2000
Policy 3	Tax to reduce to 90Q
Policy 4	Tax to reduce to 70Q

TABLE 9

Percentage Change From Base Case Policy

	1	2	3	4
Input Prices				
Capital	–	–	–	–
Labor	-0.68	-2	-5	-10
Energy	16	30	108	217
Output Prices				
Agriculture and Construction	1	1.3	2	5
Manufacturing	0.08	0	1	6
Transportation	3	4	8	13
Services	-0.3	-1	0.6	2
Energy				
Actual	0.2	11	88	187
Effective	16	30	108	217

imagine it is an effect on GNP of a <u>reduction</u> of (3 1/2 - 1) x 5 = 12 1/2%. But there will also be a reduction of investment and according to Jorgenson and Hudson's calculations, there will be a reduction of GNP by 27% relative to no increase in fuel prices.

These considerations and scenarios may all be irrelevant to the rest of the world. In all the scenarios nuclear power can be replaced by coal--to the extent it can be produced fast enough--and the cost increase is not great. But for other countries this is not true. Japan, for example, has little coal, and the transportation would quadruple the price relative to the Montana price, and perhaps <u>double</u> the delivered energy cost.

REFERENCES

1. Hogan, William W., 1979, "Dimensions of Energy Demand," Discussion Paper Series E-79-02, Energy and Environmental Policy Center, Harvard University.
2. Brooks, Harvey and Hollander, J.M., 1979, "United States Energy Alternatives to 2010 and Beyond: the CONAES Study," Ann. Revs. of Energy, <u>4</u>, 1.
3. Jorgenson, D.W. and Hudson, E.A., 1978, "The Economic Impact of Policies to Reduce U.S. Energy Growth," Resources and Energy, <u>1</u>, 205.

FAST BREEDER REACTORS - LECTURE 1

W. Marshall[*] and L.M. Davies[+]

UKAEA[*]
11 Charles II Street
London, SW1Y 4QP, U.K.

AERE Harwell, Didcot[+]

INTRODUCTION

The subject of these lectures will be the Fast Breeder Reactor. At the outset let us make some statements which will serve to illustrate the importance of this subject.

(a) The world's thermal energy reserves from fossil fuels are about 83×10^{21} thermal joules according to King Hubbert.

(b) The world's uranium reserves are about 10^7 tonnes and in thermal reactors this would yield about 4×10^{21} J of heat.

(c) Fast breeder reactors could extract about 2×10^{23} J of heat (i.e. greater than that from the total fossil reserves) from this 10^7 tonnes of uranium.

Thus fast breeder reactors can make a significant contribution to the world's future energy requirements.

In these lectures we will describe the development and construction of fast breeder reactors, their safety, the fuel cycle and the economics of the system. Lastly, we will consider the concerns that have been expressed with regard to the introduction of fast breeder reactors on a commercial scale.

FAST REACTOR

It is not our intention to delve into the physics of fast breeder reactors. If you wish to pursue this aspect then an appropriate reference (Ref. 1) has been provided in the lecture notes. However it is worth including some comments because part of the terminology is derived from it.

At this stage it is worth reminding ourselves that the term "fast reactor" does not in itself define a specific style of design any more than does the term "thermal reactor". It just means that the reactor is constructed without a moderator and consequently that the average neutron velocity is very much higher than in the thermal reactor. Typically, in a fast reactor, the most numerous population of neutrons will be in the energy range 0.2 to 0.8 MeV. In a thermal reactor, the neutron population will be roughly equally divided above and below about 1 eV. (See Figure 1).

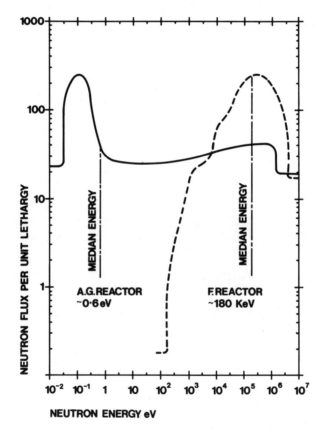

Fig. 1.   Neutron flux per unit lethargy plotted against neutron energy for a "thermal" reactor and a "fast" reactor

When the fission of uranium nuclei had been discovered, and it had been observed that a fission caused by one neutron generated two or three new ones, the possibility of a chain reaction generating power was apparent.

There was difficulty in achieving a chain reaction. Of the two main isotopes present in naturally occurring uranium, the "fissile" one, ^{235}U, in which fission can be caused by neutrons of any energy, is very rare. The other, ^{238}U, can be fissioned only by high energy, or "fast" neutrons, above about 1 MeV, but it "absorbs" neutrons of all but the lowest energies, (below about 0.1 MeV).

Therefore, for a chain reaction, the energy of the fission neutrons has to be round about thermal levels, in which case natural uranium can be used, or the proportion of ^{235}U or some other fissile isotope has to be increased substantially. Both of these routes were followed from the early days; the first led to the development of "thermal" reactors, and the second to "fast" reactors.

Let us now consider what happens to neutrons captured in ^{238}U. The ^{239}U nucleus formed decays as follows:

$$^{239}U \xrightarrow[\text{23.5 m}]{\beta^-} {}^{239}Np \xrightarrow[\text{23.5 d}]{\beta^-} {}^{239}Pu \xrightarrow[\text{24360 y}]{\alpha} {}^{235}U \text{ etc.}$$

The times are the half-lives of the decay processes. As far as reactor operation is concerned, ^{239}Pu is the end product of the chain.

The nuclear properties of ^{239}Pu are similar to those of ^{235}U, and in particular it is "fissile". So we have a way of converting the "fertile" isotope ^{238}U into "fissile" material. The quantity of fissile material which can be potentially generated depends on the material in which the fissions take place and the energy of the neutrons causing them. The important parameter is η, the number of fission neutrons generated per neutron absorbed. Figure 2 shows the variation of η with energy for two fissile isotopes. Of these η neutrons, one is needed to maintain the chain reaction. Some of the rest are lost because they diffuse out of the reactor or are absorbed in other materials such as structure or coolant. The remainder are available for capture in fertile material to turn it into fissile. If L neutrons are lost and C captured in fertile material, we have therefore, roughly

$$C \stackrel{\triangle}{=} \eta - 1 - L$$

If $C > 1$, more fissile atoms are created than are consumed. C is known as the "Breeding Ratio", and the reactor is a "Breeder" reactor. If $C < 1$, as it is in most thermal reactors, it is called the "Conversion Ratio".

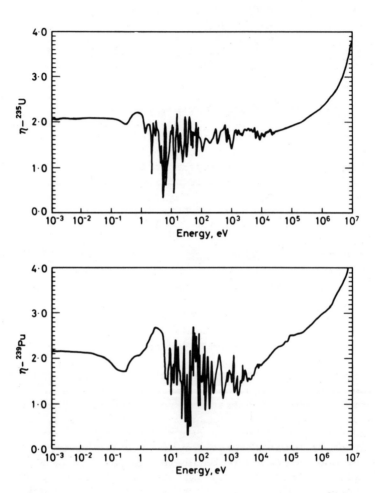

Fig. 2.   Neutron yield (η) per neutron absorbed for ^{235}U and ^{239}Pu (REF. 1)

In practice, L cannot be reduced below 0.2, so breeding is possible only for $\eta > 2.2$. Figure 2 shows how this can be brought about. A fast reactor using either of the two fissile materials will breed only if the neutron energy is above 1 MeV or so. In both cases the higher the neutron energy, the higher the breeding ratio.

Thermal reactors using ^{235}U have "conversion ratios" in the range 0.6 (LWRs) to 0.8 (heavy water and gas-cooled reactors).

If N atoms of ^{235}U are fissioned in a reactor with breeding or conversion ratio C, then CN ^{239}Pu atoms are generated. If these in turn are fissioned in a similar reactor C^2N ^{239}Pu atoms are formed (assuming for illustration that C is unchanged) and so on. The total number of atoms which can be fissioned in this way is $N(1 + C + C^2 + \ldots\ldots\ldots)$. If $C < 1$, the series converges to $N/(1 - C)$. If the fuel is natural uranium, N is 0.7% of the total number of uranium atoms supplied, and the maximum number of atoms which can be fissioned in a thermal reactor with $C = 0.7$ is, $0.7/(1 - 0.7) \simeq 2\%$ of the uranium supplied. For the PWR the value is 0.6% because so much of the ^{235}U is rejected in the initial enrichment.

For a breeder reactor with $C > 1$ the series diverges and, in principle, all the uranium supplied can be fissioned, the ^{235}U directly and the ^{238}U by converting it into ^{239}Pu. In practice, some ^{239}Pu is not immediately recoverable during fuel reprocessing, and some by conversion to higher isotopes, and the limit is about 50% of the total uranium feed which can be fissioned. Thus, in round numbers, breeder reactors can extract about 80 times as much energy from uranium as pressurized water reactors.

So, breeding is basically the conversion of ^{238}U into ^{239}Pu by neutron capture, but there are complications. ^{239}Pu is the beginning, not the end of a chain. It in turn, undergoes neutron capture and is converted into ^{240}Pu. This is a fertile isotope with properties rather like ^{238}U; it is fissile to fast neutrons, but captures neutrons of all energies. When it captures a neutron it is converted into ^{241}Pu, which is fissile like ^{239}Pu, except that it has a higher fission cross-section, especially for neutrons with energy below 0.1 MeV. ^{241}Pu may in turn capture a neutron to become ^{242}Pu, which again is fissile only to fast neutrons. ^{242}Pu also captures neutrons, and the resulting ^{243}Pu undergoes β-decay to ^{243}Am. Still higher isotopes are formed, but the quantities are small and they can be neglected for the purposes of this lecture.

DESCRIPTION OF FAST REACTOR

In this section we will describe the fast breeder reactor in outline and we will show that a variety of designs is possible.

The core of the reactor contains fuel sufficiently enriched to achieve a chain reaction. This core is surrounded by a "blanket" of fertile ^{238}U which can capture neutrons escaping from the core. In the operation of such a reactor the inventory of fissile material decreases in the core and increases in the blanket; overall it falls. The fission product inventory increases.

The core fuel is removed after a period of operation to:

(i)   Remove the fission products accumulated in the fuel. They absorb neutrons and reduce the breeding ratio of the reactor.

(ii) Adjust the fissile loading to maintain the chain reaction.

(iii) Reconstitute the fuel into new fuel because there is an irradiation limit to the operation of the fuel.

At longer intervals the blanket elements are also reprocessed to remove fission products and to separate out the plutonium for use in fabricating new core fuel. New blanket elements are made from purified ^{238}U and from new ^{238}U feed material.

Since the reactor can produce more plutonium than it consumes, this excess can be accumulated until it is sufficient to fuel a second fast reactor. Alternatively, if no system expansion is required, the ^{238}U blanket could be reduced until only enough plutonium is generated to replace that which is consumed in the core. More probably the breeding could be adjusted so that the speed with which the fuel is cycled could be relaxed so that the system as a whole is "self-sustaining" rather than "breeding".

When system expansion is required the obvious approach is to minimize the quantity of plutonium involved in the whole reactor and processing system; in that way it takes less time to accumulate enough plutonium to start up a second one.

We have already indicated that a number of variations are possible in the design of a fast breeder reactor with respect to:

(i)   the size of the 'specific plutonium inventory' - the amount of fissile material needed for a reactor to produce a given power output at a given load factor. This inventory comprises two parts - fissile material present within the reactor itself and fissile material outside the reactor in the fuel processing cycle. The former part depends, particularly, on the fuel rating and enrichment. The latter amount is dependent on the rate at which fuel passes through the reactor (i.e. burn-up and rating) and also the time taken in storage before reprocessing the fuel, the

reprocessing time itself and the time to fabricate fresh
fuel and to transport it into the reactor.

(ii)   Its 'breeding gain' (i.e. its excess plutonium production)
        and hence its 'linear doubling time' (the time taken for a
        fast reactor to produce enough plutonium to provide the
        total, i.e., in-pile and out-of-pile, inventory required by
        a new reactor. It is proportional to total plutonium
        inventory and inversely proportional to its net plutonium
        production, the latter depending on breeding gain, the
        amount of plutonium not immediately recovered from process
        residues and load factor).

    In addition to these two major factors are the practical con-
siderations arising directly from reactor design.

    Since the reactor has a liquid metal coolant the core must be
small (a few cubic metres); neutron leakage from the core into the
blanket is therefore relatively high. It follows that for neutronic
reasons a high concentration of fissile material is required for
criticality to be achieved. Thus a fast reactor requires fuel with
a high content fissile component - and such fuel is expensive
besides being a scarce resource. To produce electricity costing
about the same as that produced by thermal reactors means that the
'specific power' (in MW/t of heavy atoms) of the fast reactor must
be at least an order of magnitude greater than thermal reactors to
offset the increased fuel costs. That in turn makes it necessary
to maximize the heat transfer area of the fuel. Because the fuel
is usually in the form of cylindrical rods ('pins') an increase
in heat transfer area leads to reducing the diameter of fuel,
thereby implying an increase in length or number of fuel pins.
The cost of manufacture considerably increases as the diameter
decreases. All this demands an optimization of reactor design
to give best performance within the technical restrictions we
shall be discussing later. Depending upon circumstances that
optimization process can lead to a wide variety of design parameters.
There is therefore no unique "fast reactor design".

    The criteria applied to the choice of coolant are:

  (i)   It must be chemically compatible with the core and heat
        exchanger materials.

 (ii)   It must be commercially available.

(iii)   The pumping power required to remove the heat must be a
        minimum.

 (iv)   It must be acceptable in terms of reactor safety.

Table 1.   Variation of Design Parameters for Fast Reactors

	DESIGN 1	DESIGN 2	DESIGN 3 Reference	DESIGN 4	DESIGN 5	DESIGN 6
PLUTONIUM INVENTORIES kg Pu E 239/GWe						
In-pile inventory at equilibrium	2940	2350	2230	2230	2060	2640
Fuel cycle inventory at 75% load factor	2334	1228	1184	1042	1628	1368
Total inventory	5274	3578	3414	3272	3688	4008
PLUTONIUM FLOWS kg Pu E 239/GWe yr						
Input	1921	2050	1936	1739	2735	2160
Core	-174	-254	-231	-306	-282	-99
Blanket	+347	+404	+421	+434	+460	+405
Output	2094	2200	2126	1867	2913	2466
Net production	+173	+150	+190	+128	+178	+306
After reprocessing (Allows for some Pu held in process plant residues)	+131	+107	+147	+91	+120	+293
Equilibrium doubling time at 75% load factor years	53·4	44·8	30·9	47·9	41·2	18·2

DESIGN 1 assumes 18 months out-of-pile time, 2 fuel cycle batches, a core height of 1·2m and a mass rating of
    229 M W (t) tonnes of heavy atoms
DESIGN 2 includes an increase in the number of fuel cycle batches, a decrease in the time the plutonium spends
    in the cycle and a decrease in fuel pellet diameter compared with Design 1
DESIGN 3 is Design 2 with decreased core height
DESIGN 4 is Design 3 with decreased core fuel density and increased core burnup
DESIGN 5 is Design 3 with decreased core fuel density and decreased core burnup
DESIGN 6 is Design 5 with decreased mass rating, decreased can thickness, increased axial breeder height
    and reduced plutonium held in residues

(v)   It must be suitable to reactor core conditions (i.e. not a
      neutron moderator).

Sodium is used because it allows the use of higher power
density reactors with less pumping power. It also has the advantage
of requiring no circuit pressurisation. As sodium becomes slightly
radioactive in the primary circuit it is prudent to interpose an
additional sodium circuit and this, in turn, exchanges heat with
water. This additional circuit and the capital cost penalties
consequent on the use of sodium carry the immediate implication
that the capital cost of the fast breeder reactor will be higher
than that of a thermal reactor even though there are capital gains
from the smaller core and the use of a low pressure primary system.
However, overall, the capital cost will be higher.

Fuel and core temperatures are high - leading to enhanced
damage to the fuel during operation. However, high burn-up of
the fuel is necessary to minimize inventory losses. Metallic
fuel has achieved high burn-up but is metallurgically unstable,
particularly at high plutonium contents, so ceramic fuel (usually
oxide) is universally preferred. However the introduction of
oxygen into the fuel increases the degree of moderation which leads
to reduced breeding gain. It is possible that carbide fuels would
be better than either but not enough is known about them to plan
on their early use.

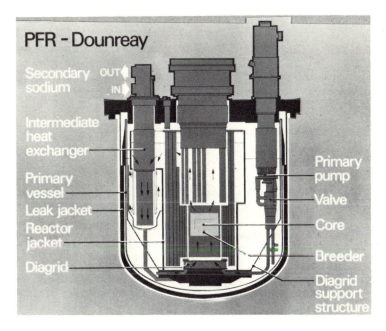

Fig. 3.   Schematic illustration of the 'internals' of a fast
          breeder core region

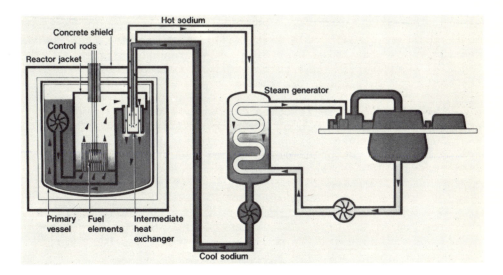

Fig. 4.   Schematic illustration of the general arrangement of a
          fast breeder reactor

Because the fuel releases fission products it has to be 'clad' in a 'can'. The cladding material chosen has not only to be compatible with the fuel and coolant but also has to withstand the rigours of service - neutron irradiation and stress. However such materials, usually steels, are neutron absorbers so the thickness of the cladding has to be minimized but still be able to withstand rigorous service.

The effect of variations in design parameters on the plutonium inventories, plutonium flows and doubling time are shown in Table 1, where it can be seen that variation in core and blanket design leads to variations in both plutonium production and in doubling time.

Figure 3 shows, schematically, the core region of a typical fast breeder reactor. The fuel elements are made of mixed plutonium and uranium oxide clad in stainless steel. The core is surrounded by a region called the "breeder or blanket", which is filled with ^{235}U, clad in stainless steel.

Figure 4 shows, schematically, the general arrangement of a typical fast breeder reactor. The heat generated from the fission process is carried away from the core by molten sodium. The hot sodium is pumped through heat exchangers so that its heat is exchanged with the sodium in the secondary circuit and this, in turn, exchanges heat with water in the secondary heat exchangers which produce steam to drive the turbines to produce electricity.

DEVELOPMENT OF FAST BREEDER REACTORS

Before about 1960 a high breeding ratio was thought to be the most important quality of a fast reactor. This meant that the mean neutron energy had to be kept high, so extraneous materials, especially moderators, had to be excluded from the core. As a result the early reactors had metal fuel.

The cores of these reactors were small; for high power operation they had to be cooled with high density coolants. Water was precluded because it is a moderator, so liquid metals - either sodium or sodium potassium alloy - were used. Many neutrons leaked from the small cores, and they were captured in surrounding regions of the natural uranium known variously as breeders or blankets.

Low power experimental reactors of this type were built in the late 1940s and early 1950s. EBR-I (USA) was the first step towards power reactors. It produced 1.2 MW and was the first nuclear reactor of any type to generate electricity.

For full scale power production the core had to be enlarged to include more fuel to permit extra power generation and to allow for the extra coolant flow. The result was EBR-II and EFFBR in the USA and DFR in the UK. When they were constructed these were seen as prototype power station reactors and EBR-II and DFR were used mainly to test oxide fuel for the next generation of reactors.

About 1960 it became clear that there was more to a fast reactor than good breeding. The cost of the fuel itself, and of fabricating it and reprocessing it after irradiation to remove the fission products and replenishing the fissile material in the core, is significant. The longer the fuel can stay in the reactor before it has to be reprocessed, the lower the cost.

The extent of irradiation that fuel undergoes can be defined in terms of the fraction of all the uranium and plutonium atoms (i.e. all the heavy atoms) which are fissioned. This is called the "burn-up". It became clear that metal fuel could only withstand modest burn-ups before the fuel-clad reaction was excessive so that it had to be reprocessed. The fuel costs associated with such frequent reprocessing were excessive, so an alternative was found in the form of oxides, either UO_2 or PuO_2 or a mixture of the two. Oxide fuel can stand 10% burn-up or more, and so has lower reprocessing costs.

Oxide fuel has other advantages over metal, mainly that it can be operated at high temperatures. The main disadvantage is that the introduction of oxygen into the core reduces the mean neutron energy, and therefore reduces the breeding ratio.

Carbide fuel (UC and PuC) should be better than oxide in many respects, but suffers the disadvantage that it is not so well understood as oxide, which is widely used in thermal reactors as well as fast. The need for reliability often dictates a conservative approach, so oxide fuel is currently used almost universally. In the future, however, it is quite possible that carbide will be preferred.

Mixed oxide fuel, a stainless steel structure, and a liquid sodium coolant have become accepted widely as the route for the development of fast breeder reactors. The use of these materials restricts the design so that all current reactors, from whatever country, show marked similarities. These are all prototypes to be followed by production reactors for the commercial generation of electricity. The developments in countries having major breeder programmes are shown in Figure 5. It is anticipated that the start-up of the first commercial size demonstration fast breeder reactor will be in France in 1982.

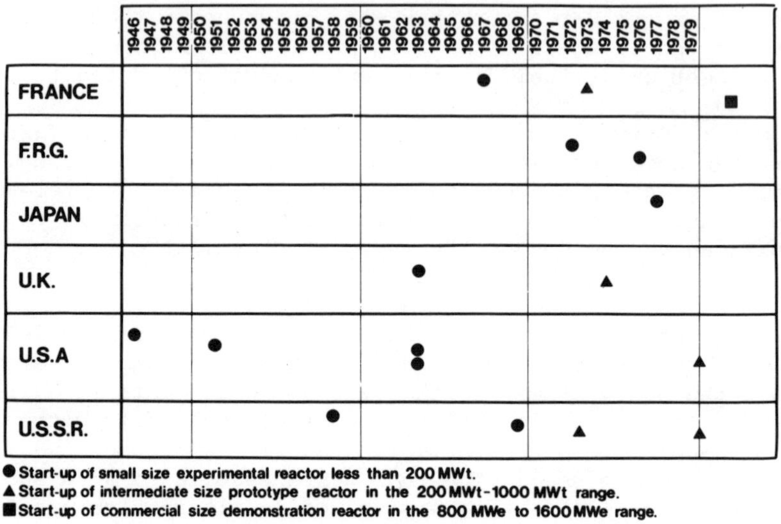

	1946	1947	1948	1949	1950	1951	1952	1953	1954	1955	1956	1957	1958	1959	1960	1961	1962	1963	1964	1965	1966	1967	1968	1969	1970	1971	1972	1973	1974	1975	1976	1977	1978	1979	
FRANCE																						●						▲							■
F.R.G.																											●		●						
JAPAN																															●				
U.K.																		●											▲						
U.S.A	●			●														● ●																▲	
U.S.S.R.												●										●					▲						▲		

● Start-up of small size experimental reactor less than 200 MWt.
▲ Start-up of intermediate size prototype reactor in the 200 MWt–1000 MWt range.
■ Start-up of commercial size demonstration reactor in the 800 MWe to 1600 MWe range.

Fig. 5.   Fast breeder reactor programmes

SUMMARY OF FIRST LECTURE

    In this lecture we have covered the nuclear mechanism of the
fast breeder reactor. The reactor is termed "fast" because
neutrons generated from fission are not slowed down ("moderated").
Plutonium is incinerated and some is produced, in the core of the
reactor. Neutrons escaping from the core are absorbed in the blanket
and in that absorption process, ^{238}U is converted to ^{239}Pu. By
design it is possible to arrange that the production of plutonium
in the blanket exceeds the incineration of plutonium in the core
and the reactor can then be called a "breeder" because, once started,
it is able to make its own fuel and perpetuate itself, i.e. "breed"
indefinitely provided it is fed with enough depleted ^{238}U - which
is in plentiful supply.

    We have seen that the technology of the fast breeder reactor
has now progressed to the point where commercial scale reactors
can now be built.

REFERENCE

Lamarsh, J.R., "Introduction to Nuclear Engineering", 1975, Addison-
    Wesley Publishing Company.

FAST BREEDER REACTORS - LECTURE 2

W. Marshall[*] and L.M. Davies[+]

UKAEA[*]
11 Charles II Street
London, SWIY 4QP, U.K.

AERE Harwell, Didcot[+]

INTRODUCTION

While various possibilities for the design of the fast breeder reactor exist, you will remember that the main development of the technology has been along the line of mixed oxide fuel, steel cladding and liquid metal coolants. We will restrict ourselves to the consideration of the design of this sort of reactor in this lecture, where we will describe some of the features in some detail. The performance of the fast breeder reactor can be varied with respect to the size of its specific plutonium inventory (the amount of fissile material required for the cycle of operation) and also the breeding gain (i.e. its excess plutonium production) and hence its doubling time, by altering some of the basic core parameters.

In general, specific inventory decreases with:

  (i)   Increasing fuel burn-up (mainly due to decreasing the ratio of the out-of-pile time to the in-pile time).

 (ii)   Increasing mass rating (MW/tonne heavy atoms) - by either decreasing pin diameter at constant linear rating or by increasing linear rating at constant pin diameter.

(iii)   Decreasing core height.

 (iv)   Decreasing axial breeder height.

  (v)   Increasing the number of batches in the fuel cycle.

 (vi)   Decreasing fuel density.

(vii)   Decreasing out-of-pile processing time.

(viii) Decreasing fuel pin can thickness.

In general, breeding gain increases with:

(i)   Decreasing burn-up.

(ii)   Decreasing mass rating.

(iii) Decreasing core height.

(iv)   Increasing axial breeder height.

(v)   Increasing the number of batches in the fuel cycle.

(vi)   Increasing fuel density.

(vii) Decreasing fuel pin can thickness.

We can add that nett plutonium production, which depends upon breeding gain, increases with decreasing amounts of plutonium not immediately recoverable from the fuel cycle during fabrication and reprocessing. In addition, the specific inventory and breeding gain decrease with decreasing load factor (i.e. availability of plant).

It is also worth remembering that because a decrease in specific inventory and an increase in breeding gain both lead to

Table 2.   Typical core parameters for an early 1250 MW(e) commercial reactor

Number of batches in fuel cycle	-	6
Fuel pellet diameter	mm	5
Can thickness/pellet diameter	-	0.075
Linear rating	W/mm	50
Mean mass rating	MW/tha	291
Max core burnup	% ha	10
Fuel density	% theoretical	80
Core height	m	1
Core diameter	m	2.66
Axial breeder height	m	2X0.4
Radial breeder thickness	m	0.4
Total sodium area/total core area	-	.410
Fuel feed enrichment	% PuE239	17.0
In-pile inventory at equilibrium	tPuE239/GWe	2.23
Out-of-pile inventory at 75% load factor and nine months Pu out-of-pile time	tPuE239/GWe	1.18
Breeding gain	-	0.209
Nett plutonium production with 2% Pu in residues	tPuE239/GWey	0.147
Equilibrium linear doubling time at 75% LF	y	30.9

a reduction in doubling time the search for an optimum value of
doubling time involves a balance of core parameters that often
work in opposite directions. Overshadowing the optimisation of
the reactor core itself are the adverse effects of long plutonium
hold-up times and plutonium retained in residues from the reprocessing plants. As a result of optimisation studies, typical core
parameters for an early 1250 MW(e) commercial reactor are listed
in Table 2.

These optimised values have an immediate implication on the
requirements of the fast breeder reactor system, both in terms of
design and its operation. We will examine some of these features
in the following sections.

HEAT TRANSFER

The specific inventory consists of two parts - the fissile
material in the reactor and that outside the reactor in the
reprocessing cycle. The amount of fissile material present within
the reactor itself depends on the heat rating and burn-up achieved
in the fuel in the reactor. Thus the higher the rating of the
fuel, the lower the specific inventory. So there is an incentive
to maximise the fuel rating.

However, there are two limitations on the design of the core
In the first instance, the cladding must not exceed $700^{\circ}C$ to
ensure the cladding will remain intact because of a general loss
of strength and, secondly the maximum temperature for the fuel
is set at about the melting point of the fuel $(2800^{\circ}C)$.

The centre temperature of the fuel during operation is related
to the linear heat rating of the fuel pin. Thus to avoid melting
of the fuel a maximum limit is set for the linear heat rating,
and the value is usually about 50 kWm^{-1}. Because there are axial
and radial variations, the average value for linear heat rating
for the fuel is about 30 kWm^{-1}. Thus a 2500 MW(t) fast breeder
reactor requires 8×10^{4} m total fuel element length.

To reduce the investment in fuel there is therefore an
incentive to make the diameter of the fuel elements as small as
possible. However, the route for manufacturing fuel elements is
complex and also the cost increases with smaller diameters and
a compromise of about 6 mm is chosen. A fuel pin diameter of
6 mm implies a total of 28 tonnes of fuel in the core of a
2500 MW(t) fast reactor.

The high heat flux can be transferred with only a small
temperature difference between cladding and the sodium coolant.

To allow for small variations in power density, temperature variations etc., the maximum coolant temperature is controlled at less than $600^{\circ}C$, which then meets the mechanical properties requirements of components.

Because the maximum coolant temperature has been "fixed" and because the high temperature differences increase stresses in the structure, the resulting design is a compromise. A typical 2500 MW (thermal) reactor would have a coolant temperature rise of about $150 - 170^{\circ}C$, a mass flow rate of 1200 Kgs^{-1} at a maximum speed of 10 ms^{-1}, and a core about 1 m high and 2.2 m in diameter.

FUEL

The fuel is usually in the form of a mixture of uranium and plutonium dioxides. Oxide is used because it is stable at high temperatures, is compatible with the stainless steel cladding, and consequently it is the fuel that has been most widely studied for use in fast breeder reactor systems. There is a possibility that the carbide form of uranium and plutonium could be used in the future because it has a higher thermal conductivity, higher density and a lower moderating effect on the neutrons.

The fuel element is a stainless steel tube which contains the fuel. Usually the fuel is in the form of sintered pellets which are made from UO_2 and PuO_2 powders, which are mixed in the correct ratio with a binder, formed into shape, sintered and ground to size. There are variations in the manufacturing of the fuel. One alternative to using sintered pellets is to fill the tube with sintered particles of oxide and then to vibrate the tube until the fuel is compacted. To obtain a high packing fraction, particles of various sizes are used. This "vibro-compacted" fuel is attractive because of its simplicity and possible cheapness of manufacture. There are other variations being considered.

On irradiation the appearance of the fuel is completely changed. Figure 6 shows a polished and etched cross-section of fuel after irradiation. You will observe that a hole has developed in the centre which is surrounded by a high density region of fuel of long "columnar" grains lying along the radii of the cylinder. This is surrounded by a region where the grains are larger than in the original material, but arranged at random and only in the outermost region is the original structure preserved.

This restructuring takes place because at intermediate temperatures in the range $1250 - 1700^{\circ}C$ the sintering process started during manufacture, continues and grains coalesce and

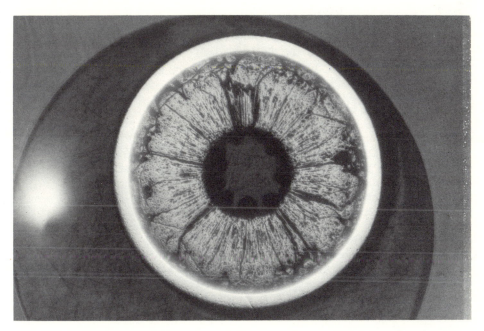

Fig. 6.   Cross-section of irradiated fast reactor fuel showing
           central hole, macroscopic cracks, columnar and equiaxed
           grain region

grow. At higher temperatures the pores migrate up the temperature
gradient because fuel atoms move from the hot to the cold side.
The pores move to the centre where they form a hole, while the
fuel recrystallises in the wake of a pore as the single crystal
or columnar grain. The columnar region is being recrystallised
as first the original pores and later bubbles of fission product
gas migrate to the central void. In figure 6 you will also see
cracks which were not present when the reactor was operating.
When the fuel is at temperature, the centre expands more than the
circumference because it is hotter, and the cracks, tapering
towards the centre, are formed. In the central region these soon
heal as the fuel is recrystallised. When the reactor is shut
down the central fuel contracts and new cracks, tapering towards
the circumference, are formed. These cracks heal on reoperation
of the fuel under the irradiation conditions.

     Fission in the fuel itself causes swelling of the fuel
because one fissile atom is replaced by two fission product atoms.
Some of the fission products are lost so that the actual volume
increases about 0.8 times the burn-up of the fuel. If this
swelling were accommodated entirely by straining the cladding,
it would cause about 3% linear strain at 10% burn-up.

Of the fission products that are produced in the fuel, about 12% are gaseous. These are Xenon and Krypton. As their solubility in the fuel is very low, they tend to be precipitated in the form of small bubbles which eventually link up at grain boundaries to form channels by which most of the gas is released. The quantity is very large. At 400°C and 1 atmosphere pressure the gas generated by 10% burn-up of fuel with a density of 10^4 kgm^{-3} occupies 53 times the fuel volume. This gas is accommodated in an empty volume or "plenum" at one end of the fuel element.

It is possible to make the fuel more porous so that this swelling can be accommodated by the elimination of pores. Another possibility is that the pellets can be made with a central hole for the same purpose.

A wide range of elements are produced as fission products, forming a very complex chemical system in the fuel. Also the temperature difference that exists between the surface and the centre of the fuel further complicates the chemical system formed. Probably the dominant chemical factor is the oxygen potential of the fuel which changes as the fissile atoms are replaced by fission product atoms.

The purpose of the cladding is to maintain the configuration of the fuel and to prevent radioactive materials getting into the coolant. A small amount of the cladding is removed by the coolant and a fraction of the corrosion products will become radioactive but the amount is tolerable. We have already seen that as the fuel swells, it releases fission product gases and some of the fission products tend to corrode the cladding. The fuel and cladding interact mechanically and this is a very complex process. The fuel swells and presses onto the cladding at the start of irradiation but the stresses are limited by creep deformation. The cladding itself swells under fast neutron bombardment (an effect not observed in thermal reactors) and this swelling may at some stage exceed the fuel swelling rate, so that the gap between them increases. We will return to this later.

As we have mentioned before, the various mechanisms occuring during irradiation of the fuel interact in a very complex way, so that the main guarantee of fuel behaviour for the design of a fast breeder reactor is provided by extensive irradiation testing of such fuels. For such evaluation there is the requirement for test facilities - both for irradiation and post irradiation examination.

The fuel is in the form of sintered pellets. Above and below the core fuel are axial breeder regions consisting of natural or depleted UO_2. The volume of the fission product gas plenum is about equal to that of the fuel so that the pressure in the fuel element at the end of life is some 5 or 6 MPa.

CLADDING AND STRUCTURAL MATERIALS

Let us now say something about the effect of neutron irradiation on the cladding and structural material.

It has been calculated that each and every atom in the iron
and steel structure is displaced from its lattice site about 70
times on average during neutron irradiation to full lifetime in
the fast breeder reactor. This neutron irradiation has an effect
on the properties of these materials. The main effects are
swelling, creep deformation, embrittlement and hardening.

Each neutron scattering event produces a cloud of inter-
stitial atoms and vacancies. These diffuse through the material
and normally would recombine in due course. In addition, the
irradiation produces helium atoms by (n,α) reactions. As they
diffuse, the vacancies tend to form clusters around helium
nuclei while interstitials accumulate at grain boundaries.

The clusters of vacancies form minute voids, and these grow
causing the material to swell. The swelling depends strongly on
temperature and varies widely for different materials. In extreme
cases, volume increases of about 10% are possible. (See Figures
7 and 8).

Creep deformation is enhanced by neutron irradiation and
this is important. In sub-assembly deformation the hexagonal
wrapper tries to become a cylinder.

The defects produced by neutron irradiation reduce the
mobility of dislocations in the material and so inhibit plastic
strain. This results in an increase of the yield stress. At
higher irradiation temperatures however the defects tend to be
annealed out by a mutual annihilation of interstitials and
vacancies. So that the mechanical properties are closer to those
of the unirradiated material.

At even higher irradiation temperatures (above about 700°C)
the helium tends to diffuse to grain boundaries where it collects
as small bubbles and these cause a loss of cohesion so that the
grains can be torn apart more readily. The result is a loss of
high temperature ductility.

Sodium has a slightly corrosive effect on the austenitic
stainless steels which are used for the canning materials, where
it tends to preferentially remove chromium and nickel. However,
the removal rates are quite low. But it is dependent upon the

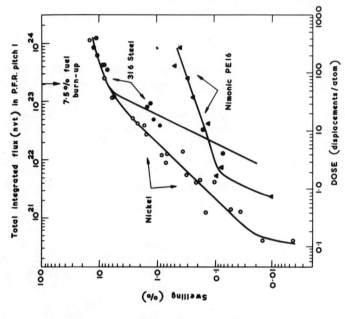

Fig. 8. Void swelling plotted against displacement dose for a variety of
materials (after J. Hudson et al.,
Conference on Foids, BNES, 1971)

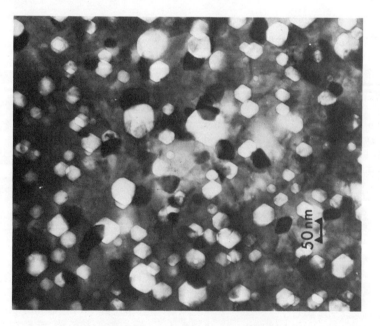

Fig. 7. Voids in 316 steel (after D. Mazey)

oxygen content of the sodium, the higher the oxygen content, the greater the corrosion. For exmaple, at $500^{\circ}C$, 316 type steel is corroded at about 2 μm per year by sodium with 10 ppm of oxygen, and at 10 μm per year with 30 ppm of oxygen.

Carbon can be transported between the different components exposed to sodium at a rate which depends on the carbon activity in the sodium. Depending on temperature carbon tends to be lost from ferritic material with high carbon activity, causing it to lose strength, and to be gained by austenitic steel, reducing its ductility.

Sodium tends to cause very high thermal stresses on structural components because it transfers heat so readily. Temperature variations or fluctuations in the coolant are transmitted to the structure and induce high surface stresses, which may cause fatigue damage.

The steels available to meet these demanding requirements fall into three classes; austenitic stainless steels, nickel alloys and ferritic alloys. Austenitic steels have relatively low yield strength but high ductility, and they resist corrosion if the oxygen content from the sodium is controlled. They suffer, however, from irradiation embrittlement, are susceptible to thermal shock and irradiation swelling. Cold work of austenitic steels tends to reduce, but does not eliminate, the swelling during fast neutron irradiation.

High nickel alloys are stronger and more resistent to irradiation swelling. The higher neutron absorption in nickel than iron is offset by the greater strength, allowing less material to be used. Alloys with a high nickel content such as Nimonic 80A (75% nickel) corrode too fast in sodium. Ferritic steels have rarely been used in reactor primary circuits because of the loss of strength and decarburisation that occurs.

CORE STRUCTURE

It will be remembered that the fuel elements are stainless steel tubes about 6 mm in diameter and 2.5 to 3 m long, containing a 1 m section of core fuel. About 80,000 of these go to make up the core of a 2500 MW (thermal) reactor, and obviously they cannot be handled one by one. They are made up into "sub-assemblies", each consisting of 200 - 300 elements in a hexagonal steel tube or wrapper. This arrangement allows the coolant flow to each sub-assembly to be regulated by means of adjustable restrictors at the inlet so that the outlet temperature is approximately uniform across the core, in spite of the variation of power density. The fuel elements are located in the sub-assemblies

either by transverse grids or by wires run helically round each
element.

Irradiation swelling leads to distortion of the sub-assemblies. The neutron flux is higher at the centre of the core than
the outside so each sub-assembly experiences a greater fluence
and therefore greater swelling on the side towards the core
centre. As a result it tends to become curved with the convex
side inwards. The extent of bowing depends on the dimensions as
well as the fluence and temperature, but if unrestrained it could
result in displacements of about 10 to 50 mm of the tops of the
sub-assemblies.

Besides using void swelling resistent material the distortion
can be avoided by rotating the sub-assemblies from time to time
to equalise the swelling, but it is more usual to prevent it.
This can be done either by a passive restraint structure, which
surrounds the core and resists any outward movement after clear-
ances between the assemblies have been taken up, or by an active
restraint mechanism consisting of clamps which can be tightened
up after the core has been assembled and slackened to allow fuel
to be changed. The restraint structure can be prevented from
swelling by keeping it cool and in a region of low neutron flux.

The fuel in the outer part of the core has a higher concentration of plutonium than the inner to compensate for the lower
neutron flux and make the power density more uniform across the
core. There may be two or more enrichment zones of this type.
The control rods occupy sub-assembly positions in the core. They
consist of neutron absorbers, usually boron carbide inserted and
withdrawn by mechanisms above the core. The core is surrounded
by the breeder, the elements of which are contained in similar
sub-assemblies or may be a separate structure. The core rests
on a support structure called the Diagrid, which serves also to
distribute coolant to the sub-assemblies.

PRIMARY COOLANT CIRCUIT

The primary coolant circuit consists of the reactor core,
intermediate heat exchangers and circulating pumps. One variation
in the design is whether the pumps should be in the same vessel
as the reactor core (a pool reactor) or in separate vessels
(a loop reactor) (see Figure 9). The pool layout has the
advantages that all the primary circuit is contained within a
vessel of simple form which has no penetration, which minimizes
stresses and is exposed only to cool sodium. The large diameter
may make handling of irradiated fuel easier because there may be
room within the reactor vessel to store it for a period to allow
the fission product heat generation to decay. The pool reactor

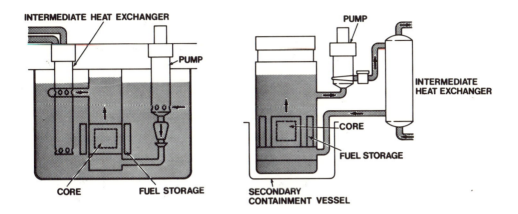

Fig. 9.   Schematic arrangement of 'Pool' and 'Loop' type fast
breeder reactors

roof is however large (possible 15 m in diameter) and highly
stressed because it carries the weight of the core, neutron
shield, primary pumps and motors and the intermediate heat
exchangers. It has to be a complex and expensive structure
compared with the roof of a loop reactor vessel.

A loop reactor vessel is smaller, part of it and of the
connecting pipework is exposed to hot sodium and part to cool,
so there are thermal stresses, and the complex shape of the
vessel may lead to stress concentrations. However, the vessel
and pipework can be inspected more easily than the stressed
members within a pool vessel. A loop reactor is more prone
to damage due to thermal shock if either the reactor or a steam
generator shuts down suddenly because of an accident.

The sodium coolant which flows through the core becomes
radioactive. It is usually thought too dangerous to use highly
radioactive sodium to raise steam, because a leak in the sodium/
water heat exchanger would allow a violent chemical reaction
which might cause enough damage to release radioactivity to the
environment. It is normal therefore to incorporate a secondary
sodium circuit which receives heat from the radioactive primary
sodium but is protected by a neutron shield from becoming
radioactive itself.

The coolant is usually circulated by pumps at the bottom of long vertical shafts. The surface of the sodium is covered with an atmosphere of argon at a pressure slightly above atmospheric. The reactor vessel and pipework have therefore to withstand only modest pressures of a few hundred kPa. They are made of welded stainless steel and the external boundary of the primary circuit is double-walled, the space between the walls being monitored to detect leaks.

STEAM PLANT

The secondary sodium coolant is used to generate steam that can be used in a turbine to drive an alternator and generate electricity. The steam cycle is very similar to that in a conventional power station so we will not dwell on that aspect.

Because of the severity of the chemical reaction between sodium and water, great care has to be taken that leaks are very unlikely in the steam plant and that, if a leak should occur, the resulting damage can be controlled. The welded joints between the tubes and tube plate have to be made to the highest standards and subjected to rigorous inspections.

If a large leak in the heat exchanger should develop suddenly the rest of the plant has to be protected by bursting discs or by some other means of relieving pressure rapidly. If there is a small leak, say due to a crack in the weld, the remedial action has to be taken before it grows or spreads to adjacent welds or tubes. Fortunately, a small leak can be detected by monitoring for the presence of hydrogen which is released from the sodium water/steam reaction. Once hydrogen is observed the heat exchanger unit concerned can be isolated and drained and the leak repaired.

If heat is transferred to the primary coolant at the rate of 2500 MW(t), the nett work output of the steam plant is 1060 MW(e) giving a thermal efficiency of 42%.

FAST BREEDER REACTORS - LECTURE 3

W. Marshall[*] and L.M. Davies[+]

UKAEA[*]
11 Charles II Street
London, SW1Y 4QP, U.K.

AERE Harwell, Didcot[+]

INTRODUCTION

In this lecture we will discuss two areas associated with fast breeder reactors. The first is safety and the second is the fuel cycle.

SAFETY

The risk caused by nuclear reactors is that the radioactive materials they contain may escape and injure members of the public in the course of an accident. Although we realise we are in the wake of the Three Mile Island incident, let us state that such reactor accidents are improbable events. It is much more likely that the events leading to such accidents will be detected and that the reactor will be safely shut down without damage. There is always the risk that multiple faults or errors will result in inappropriate corrective action being taken and that the reactor plant will be severely damaged. In a fast breeder reactor the most likely outcome of such an accident would be that the containment system would prevent significant leakage of radioactive material into the environment. Only if all the various containment systems failed in a catastrophic and sudden manner would there be a serious release of radioactive materials.

We will not be considering the consequence of releases of radioactivity in the event of these remote accidents in this lecture. However, we have included some references at the end

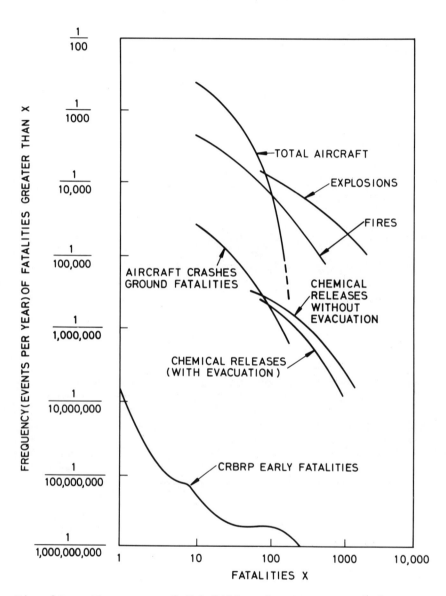

Fig. 10.   Frequence of fatalities due to man-caused events
            occurring within ten miles of the CRBRP site
            (from CRBRP-PMC 78-01)

of these lecture notes for those who wish to pursue these aspects
further.

Much attention is given to the assessment and understanding
of hypothetical accidents. Such studies identify those areas of
design and operation to which attention should be paid. So that
when a fast breeder reactor (or any other reactor for that matter)
is properly designed, properly built and properly operated the
release of radioactive materials is very unlikely indeed.

This point is illustrated in Figure 10, where the probability
per year of fatalities is plotted against the number of early
fatalities in accidents for the cases of dam failures, aircraft
accidents, thermal reactors and the Clinch River Breeder Reactor.

Let us now say something about the safety features of fast
breeder reactors.

The radioactive materials present in the fast breeder reactor
include the fuel, the fission products generated in the fuel as
burn-up proceeds, and the activation products (mainly the radioactive sodium and the products of clad corrosion).

One of the inherent safety features of the fast breeder
reactor is that there are several barriers between the radioactive materials and the external environment. There are usually
three barriers; there is the fuel element cladding, the primary
coolant containment, and the reactor building. (These are shown
schematically in Figure 11). For radioactivity to be released,
these three barriers (two in the case of ^{24}Na) have to be breached.

The second inherent safety feature of a sodium-cooled fast
breeder reactor is that the coolant pressure is low, so that the
coolant containment is lowly stressed and unlikely to fail.
However, even if this failed, it can be seen that it is quite
easy to design the system so that the core can still be cooled.
(This can be done by surrounding the reactor vessel in a hole in
the ground, so that even if the vessel and leak jacket fail, the
primary coolant still covers the reactor core and the intermediate
heat exchangers). So there are two main advantages of the low
pressure coolant; it is unlikely to fracture the containment and
even if it does, it is quite easy to remove the fission product
decay heat and thus prevent the fuel and its cladding from
becoming too hot. This feature is in contrast with the gas and
water cooled reactors where the coolant pressure is high and the
concern is providing adequate cooling in the event of coolant
containment failure.

The third inherent safety feature of the fast breeder
reactor is the large thermal inertia of the primary coolant which

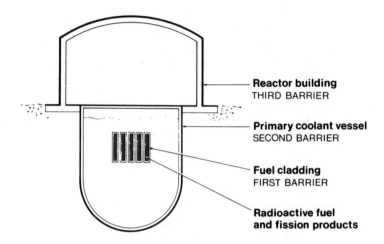

Reactor building
THIRD BARRIER

Primary coolant vessel
SECOND BARRIER

Fuel cladding
FIRST BARRIER

Radioactive fuel
and fission products

Fig. 11.   Barriers to the release of fuel and fission products
            in the event of an accident

tends to prevent overheating of the fuel and its cladding. You
will remember that the fuel cladding is more stressed at the end
of its life in the reactor because of the accumulation of fission
product gas. It will remain intact provided it is kept cool.
(Failure temperature will depend on the details of the burn-up
achieved, the design of the fuel pin, and the time at temperature
during such an excursion). These effects are illustrated in
Figure 12, which shows the rate at which the primary coolant
temperature rises when the reactor is shut down, assuming a
complete failure of the secondary cooling. You can see that
provided the primary coolant continues to circulate, the timescale
for overheating is quite long. However, there is the requirement
for an emergency cooling system to remove the "decay heat". But
the timescales involved in its operation are quite long.

    The fourth inherent safety feature is what is known as the
Doppler effect, which reduces the rate of fission in the fuel as
the temperature increases. It provides a reliable prompt negative
reactivity feed back. So that if the fast breeder reactor is
subjected to a reactivity increase for some reason, the Doppler
effect acts to limit the consequent power rise.

    Besides these inherent safety features, there are safety
disadvantages in a fast breeder reactor compared with a thermal

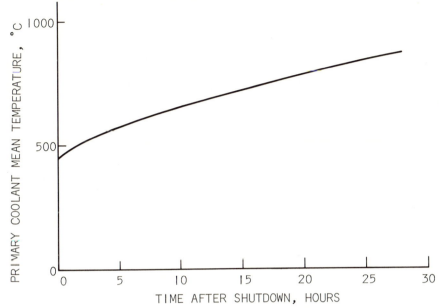

Fig. 12.   Rate of primary coolant temperature rise in a 'Pool'
            type fast reactor, assuming complete failure of
            secondary cooling

power reactor. The fuel temperature is higher, there is a higher
power density, and there is the possibility of an accidental
increase in reactivity.

     The higher fuel temperature in the presence of the sodium
coolant implies the risk of rapid vaporisation of the coolant.
If some molten metals are mixed with water the vaporisation is
sometimes violent, giving rise to the phenomenon called a "steam
explosion". Significant sodium vapour-molten oxide fuel explosions
have not been produced under realistic experimental conditions
and may be impossible, but until this is proved the possibility
has to be taken into account.

     The high power density means that if the fuel is uncooled,
its temperature rises very rapidly. If the reactor is not shut
down, and for example, if a central fuel element heats up, its
mean temperature could rise at some 600 Ks^{-1}.

     The risk of increasing reactivity arises partly from the
possibility of moving control rods accidentally, but also because
the fuel in a fast reactor is not in its most reactive configuration. For example, if it were rearranged into a more compact
shape, the reactivity would increase. However, the probability
of such an event occurring is very low.

THE FUEL CYCLE

There are several fuel cycles which involve uranium and
plutonium. Each have their own characteristics and it is important that these should be clearly understood and distinguished
one from the other. We shall distinguish four fuel cycles and
these are described in Figure 13 in an outline form.

The first fuel cycle shown in Figure 13, the once-through
fuel cycle, is that which most countries in the world are now
operating. The uranium fuel is put through a thermal reactor
and is then stored, together with its plutonium content, without
reprocessing. This is done either because reprocessing capacity
is not available or, in the case of the United States and Canada,
because it is judged to be premature and unnecessary. The
situation in the UK is different from the rest of the world
because our thermal reactors are gas-cooled and the spent fuel
from a gas-cooled reactor is not easily stored for long periods
of time. In our case, therefore the fuel must be reprocessed
for environmental reasons to separate out the plutonium, the
uranium and the fission products.

The second fuel cycle is that where the fuel is reprocessed,
the plutonium is extracted and is refabricated into fuel for the
thermal reactor. This is called thermal recycle of plutonium.
It is the fuel cycle which a number of countries are planning
to operate in the near future. The exceptions are the United
Kingdom and France, which believe it to be both unnecessary and
an inefficient use of plutonium, and the United States which
considers it to be a proliferation risk. The economic attraction
of recycling plutonium into thermal reactors is very simply
stated. It replaces the need for some fresh uranium fuel and
could be applied at a relatively early date because thermal
reactors are already available.

The third fuel cycle is that to launch the fast reactor and
uses the plutonium made in thermal reactors. In this fuel cycle
the spent fuel from a thermal reactor is reprocessed and the
plutonium is stored until it is needed. Notice that, in these
three fuel cycles, there is no urgency to get the plutonium
fabricated into fresh reactor fuel because in the once-through
fuel cycle, the plutonium is not used at all, in the case of
thermal recycle of plutonium, the value is marginal unless the
cost of uranium increases substantially and, in the third case,
fast reactors are simply not yet available to use the plutonium.

The fourth fuel cycle is that used to maintain fast reactors
in operation. That fuel cycle contains three main parts:

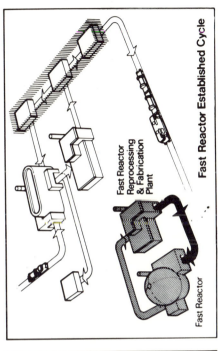

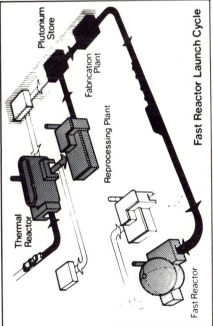

Fig. 13. Various fuel cycles

   (i)   The reprocessing of the core and breeder elements to extract the fission products from them and recover a mixture of uranium and plutonium oxides in the correct proportions to make up a fresh charge of fuel for the fast reactor core.

 (ii)   The fabrication of fresh fast reactor fuel elements from this recycled material.

(iii)   The disposal of the fission products (and actinides) as high level waste.

The waste disposal process is very similar to that used for thermal reactors so we shall say nothing about it in this lecture. However, it is important to recognise that the reprocessing of the core and breeder fuel and the refabrication into fresh fuel is very intimately linked to the operation of the system as a whole because it is essential, for good economic operation, for the plutonium to be extracted from the spent fuel and returned to the reactor within a period of say, two years, to keep the total plutonium inventory at a reasonable level. Typically, therefore, fast reactor fuel will be reprocessed, at the latest, about 12 months after it comes out of the reactor. This is in contrast to the reprocessing of fuel from water-cooled thermal reactor systems where, typically, the fuel will have been allowed to cool for five to ten years before reprocessing.

In addition to these differences of emphasis, the fast reactor reprocessing plant itself is significantly different from the reprocessing plant for thermal reactors. The percentage of plutonium in the material going through the plant is much higher than it is in thermal reactors but the total bulk of material is lower in quantity. These plants are designed with such a geometry that a criticality accident is impossible. Therefore, the overall size of a fast reactor reprocessing plant is smaller, the pipes are of a narrower diameter and the various components have a different geometry compared to a reprocessing plant for thermal reactor fuel. These differences are not of a fundamental kind but they need to be defined and the process needs to be demonstrated and fully proven.

Reprocessing of spent fast breeder reactor fuel consists of the following major operations:

  (i)   Dismantling, shearing and dissolution.

 (ii)   Separation by extraction of uranium and plutonium from fission products and trans-plutonium elements.

(iii)   Transformation of the recovered uranium and plutonium
        into useable solids.

 (iv)   Conditioning of the wastes.

     Because of the complex structure of the fast breeder reactor
fuel bundle, significant quantities of metallic sodium may be
trapped in the structural parts of the fuel bundle or in defective
fuel pins. This sodium must be completely removed before the
fuel is transferred into water or nitric acid. The residual
sodium can be removed by melting or by steam or humid CO_2 treatment. Any residual sodium can be neutralised by oxidation after
the shearing stage.

     The fuel bundles are then dismantled, the end pieces are
cut off and, depending on the structural assembly, the shroud is
removed by mechanical or laser cutting.

     Pulverisation of the sheared fuel prior to dissolution has
been considered as an optional step. In the voloxidation process,
the sheared fuel segments are dropped into a rotating calciner,
possibly equipped with crushing balls. For mixed oxides containing less than about 25% PuO_2, heating of the fuel in air or
oxygen at a nominal temperature of $450^{\circ}C$ transforms UO_2 into
U_3O_8; the resulting expansion of the lattice structure breaks
up the fuel grains releasing the tritium and a high proportion
of the fission gases.

     In general, there are two stages of dissolution in concen-
trated nitric acid:

 (i) The major part of the fuel dissolves rapidly.

(ii) A residual part, enriched in plutonium and noble metal
     fission products, dissolves slowly.

     Generally, there remains an insoluble part consisting mainly
of noble metals, fission product refractory compounds, and, in
some cases, undissolved fuel particles. If the amount of plutonium in the insoluble residue exceeds a certain value, its
recovery can be achieved by the addition of HF in a separate
dissolver.

     In the feed preparation for plutonium extraction the insol-
ubles are removed by filtration or by centrifuging. The total
quantity of insolubles to be handled is about 1% of the initial
amount of fuel and the heat released from these insolubles is of
the order of 1 W/gm. The uranium and plutonium extraction flow
sheet is shown in Figure 14. In principle, the process and
equipment used for the extraction of uranium and plutonium from
fast breeder reactor fuel are similar to those used in thermal

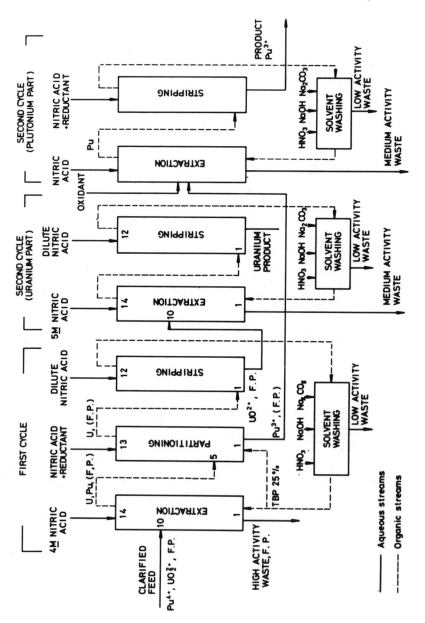

Fig. 14.   Purex Flow Sheet

reactor fuel reprocessing.

The separation of uranium and plutonium is based on the chemical reduction of the tetravalent to trivalent plutonium, this higher plutonium concentration requires the use of larger amounts of reducing chemicals, such as U^{4+}, Fe^{2+}, and $NH_2OH-N_2H_4$.

This in turn, leads to a near doubling of the volume of the solution and of the wastes. As an alternative, electrolytic reduction seems most promising but may not be sufficiently reliable for use in radioactive areas.

The method for purification of separated uranium is essentially the same as that used for LWR fuel reprocessing. The purification of plutonium differs because of the larger amount of plutonium involved and the necessity to keep the plutonium content of the wastes as low as possible.

The conversion of uranium and plutonium into oxides is the same as that for thermal reactor fuel reprocessing.

The fuel fabrication process starts with the mechanical mixing of the UO_2 and PuO_2 powders. The conventional fabrication process includes the following steps:

  (i)   Pressing green pellets.

  (ii)   Sintering the green pellets into finished pellets.

 (iii)   Grinding to size (optional).

  (iv)   Drying sintered pellets.

   (v)   Assembling sintered pellets into stacks.

  (vi)   Adding axial blanket pellets to the stack, together with other devices.

 (vii)   Inserting the pellet stacks into one-end welded tubes.

(viii)   Establishing the proper inert atmosphere inside the fuel pin.

  (ix)   Closing the pin.

   (x)   Non-destructive examination of the pin.

  (xi)   Acceptance and final identification of the pin.

 (xii)   Storage prior to assembling into bundles.

(xiii)   Bundle assembly and inspection.

 (xiv)   Bundle loading into the wrapper tube.

  (xv)   Head and foot assembly.

(xvi)   Acceptance and final identification of fuel assembly.

(xvii) Storage until shipment.

    Pelletizing and pin loading operations must be carried out in glove boxes or their equivalent, in contrast to the production of uranium LWR fuel, because all plutonium oxide dust must be contained down to very low levels. Therefore, operations are cumbersome and physical inventory checks, either approximate or highly accurate, can be made only at temporary storage stations which exist to allow time for process control and production interruptions. Control of the fissile content of the finished pins is essentially a routine operation.

    Shielding must be provided against gamma-rays and neutrons owing to the presence of many plutonium isotopes and their radio-active daughter products. In addition, force draught cooling is required because of radiation heat generated in concentrated storage areas. Shielding must not only protect the body but also the hands and arms of operators or maintenance crews. In storage areas, separation shields must be provided to prevent neutronic interaction between fissile masses.

    Tracing plutonium through the fabrication plant, along the production line as well as along the line of plutonium recovery from scrap and the measuring of plutonium left in final waste requires a variety of measuring methods and devices.

    Plutonium fuel development laboratories have now operated for two decades, while plutonium fabrication plants have operated for one decade.

REFERENCES

Kelly, G.N., Jones, J.A., and Hunt, B.W., "An estimate of the Radiological Consequences of Notional Accidental Releases of Radioactivity from a Fast Breeder Reactor", NRPB-R53, HMSO.
"Some Aspects of the safety of nuclear installations in Great Britain. Replies to questions submitted by the Secretary of State for Energy to the Nuclear Installations Inspectorate in October 1976", HMSO.
Wilson, R., "Physics of liquid metal fast breeder reactor safety", Reviews of Modern Physics, Vol. 49, No. 4, October 1977.

FAST BREEDER REACTORS - LECTURE 4

W. Marshall[*] and L.M. Davies[+]

UKAEA[*]
11 Charles II Street
London, SWIY 4QP, U.K.

AERE Harwell, Didcot[+]

INTRODUCTION

So far in these lectures we have looked at some fundamental
points concerning fast reactors, we have reviewed the technological
problems in developing them and we have examined the fuel cycle
which is required to maintain them. We shall now turn to the
more difficult task of discussing the economics of fast reactors.
This discussion must be of a different kind from that we have
given so far for a number of reasons which can be summarised as
follows:

(i)   It is notoriously difficult to predict the cost of new
      technological innovations in advance of their introduction
      on a true industrial scale. This is especially true for
      large complex projects and, as we have seen, the fast
      reactor is a more complex reactor than a thermal reactor and
      thermal reactors themselves are more complex than
      conventional power stations. The very high energy density
      in the core, the novel use of sodium as a coolant, the
      necessity to exchange heat between sodium and water, the
      unique safety features of the fast reactor and
      the intimate integration of the fuel cycle for the fast
      reactor operation all pose areas of major uncertainty.
      Furthermore, only three prototype fast reactors are
      operating in the world today - in the USSR, in France
      and in the UK. Even when the first reactor of full
      commercial size is built, that will not by itself give an
      estimate of future costs because a first full size
      demonstration reactor must necessarily cost more than

287

one produced as part of a programme of power station
production.

(ii)   So far as the cost of the reactor itself is concerned,
        it is probably fair to judge now that the performance of
        the core will be satisfactory and the major areas for
        development in the future concern the development of
        components like the sodium water steam generators. That
        involves considerations of production engineering and
        routine production of components in a style which we have
        only previously seen for light water reactors. The emphasis
        of that work must switch from nuclear R & D to practical
        production in industry.

(iii)   On the fuel cycle, each individual step has been demon-
        strated on either a laboratory or prototype scale but a
        complete and dedicated prototype fast reactor fuel cycle
        plant has not yet operated anywhere in the world and,
        obviously, that makes the prediction of the ultimate fuel
        cycle costs when performed on an industrial scale that
        much more difficult.

(iv)   The absolute economics of fast reactors do not themselves
        determine when fast reactors should be installed because
        the use of sodium and the addition of an extra intermediate
        circuit suggests that electricity produced by fast reactors
        will be more expensive than electricity produced by thermal
        reactors of the present day. The real attraction of fast
        reactors as we have seen in the earlier lectures is that,
        once launched, they require no fresh uranium ore. The
        economic installation of fast reactors, therefore, depends
        not only on the economic factors we have reviewed but upon
        the way in which the price of uranium ore varies. This is
        itself difficult to predict because it depends upon the
        law of supply and demand and both the future supply of
        and demand for uranium ore cannot be predicted with any
        certainty.

Despite these difficulties, we must find a method of discussing the economics of fast reactors in the future. Most discussions published so far have begun with a set of assumptions
about world demand for electricity and world supply of uranium
and have then attempted to deduce the variation of uranium ore
price with time and hence the "breakeven" point for fast reactor
technology. The difficulty with all those discussions is that it
is unclear how much the final conclusions are dependent upon the
initial assumptions. In this lecture, we shall therefore adopt
a different approach which has been found useful in recent international discussions to summarise and incorporate a wide range of
technical views upon the points we have discussed earlier.

The approach we will set out in the remainder of this lecture

may seem complicated or over-elaborate at first sight but we have
chosen to give it here because, in practice, it has been found
useful and the ideas are generally applicable to the introduction
of any new technology whether it be fast reactors, coal, solar
power, etc.

The factors which influence the economics of nuclear power
may be set out as follows:

Macro-economic Factors

Gross National Product (GNP).

Balance of payments (imports/exports).

Resource utilisation

- capital intensiveness
- level of innovation/risk/development and lead times
- job intensiveness
- level of qualified manpower required
- job satisfaction.

Infrastructure and environment

- infrastructure required (services/transport etc)
- impact on environment
- influence of ecological and non-proliferation considerations,
  and the sociological environment
- availability of fuel cycle services.

Micro-economic Factors

Availability and price of uranium ore.

Cost of reprocessing and waste disposal.

Cost of fabricating plutonium-bearing fuel relative to that for
uranium-bearing fuel.

Enrichment costs.

Capital cost of breeder reactors and of fuel cycles necessary to
sustain them.

Relative costs of spent fuel storage, conditioning and ultimate
disposal.

The two factors which must influence the relative evaluation
of thermal reactors and fast reactors are the cost and availability
of uranium ore to an individual country on one hand, and the
capital cost of fast breeder reactors and the fuel cycle needed
to service them on the other hand. We shall consider both these

factors as unknown parameters and we shall permit both to vary
with time. We expect the cost and availability of uranium ore to
go up with time as the uranium reserves of the world become
exhausted and we expect the capital cost of the reactor and fuel
cycle plants for fast reactors to start at an initial high value
and decrease with the benefits of scale as time progresses. We
shall use these two parameters to establish a phase diagram of
the type shown in Figures 15, 16, 17, 18, 19 and 20. In these
diagrams the horizontal axis is "the uranium (U_3O_8) price in $/lb"
and the vertical axis is "the premium over the LWR once-through
fuel cycle" which is defined as "the cost of electricity in
$/kW(e) total present worth, relative to the cost of electricity
in the same units computed for an LWR operating on the once-
through cycle with uranium priced (arbitrarily) at $25/lb".

Because we are concerned with the relative economics of
thermal and fast reactors, we must look at both the cost of maintaining fast reactors in operation and the cost of launching them.
This drives us to consider the cost of reprocessing thermal reactor
fuel to get the initial plutonium inventory which is required and
that in turn obliges us to look at the alternative uses of plutonium, namely for recycle back into thermal reactors. In brief,
therefore, we cannot examine the economics of fast reactors without looking at the full range of options for the use of plutonium.
We shall discover, therefore, that the phase diagrams describe,
in a simple form, the main features of the "plutonium economy"
and we shall find that this is helpful for our discussion of
economics in this lecture and our discussion of proliferation
questions in the last lecture.

We cannot, however, discuss everything in our relative
assessment and we shall look only at light water reactors as
examples of thermal reactors. This means that we look at:

   (i)   The once-through cycle using light water reactors.

  (ii)   Reprocessing with uranium cycle into light water reactors
        and plutonium storage.

 (iii)   Reprocessing with both uranium and plutonium recycle to
        light water reactors.

  (iv)   The launching of fast breeder reactors.

   (v)   The maintenance of fast breeder reactors once launched.

ALGEBRAIC DEFINITIONS

The algebra set out below may be performed in any of three
main sets of units, viz:

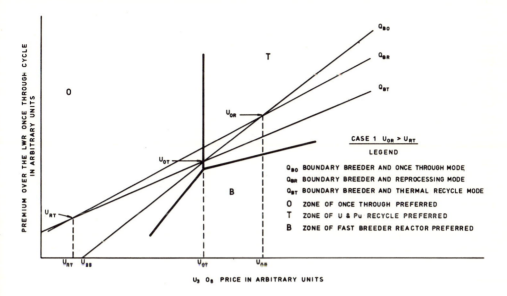

Fig. 15.

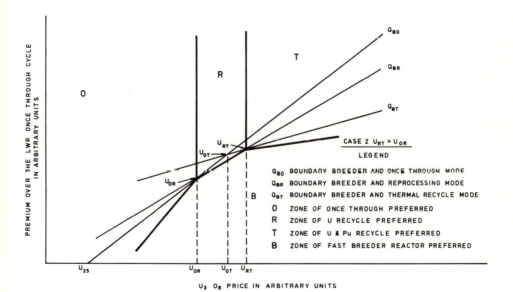

Fig. 16.

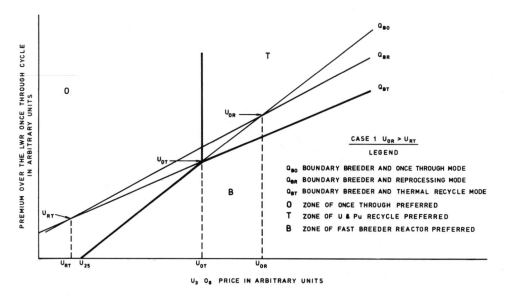

Fig. 17.

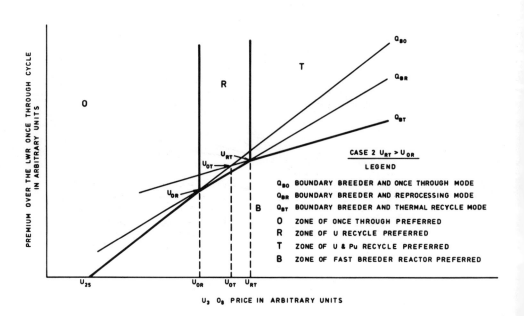

Fig. 18.

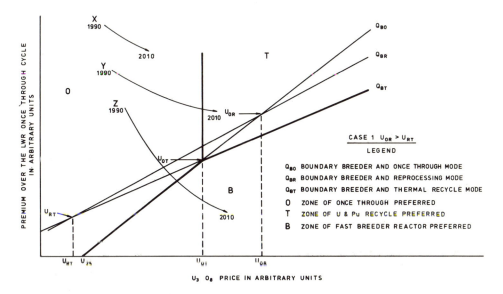

Fig. 19.

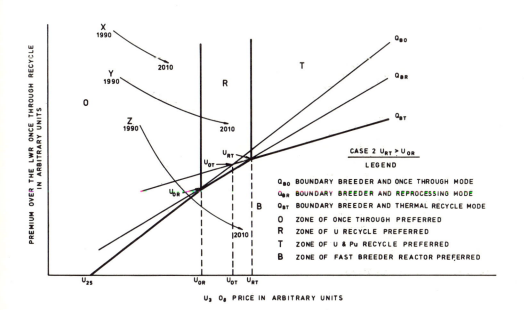

Fig. 20.

(i) $/kW total present worth at the assumed lifetime, discount factor and load factor.

(ii) $/y per kW installed capacity, related to $/kW present worth by the annuity factor A derived from the assumed lifetime and discount rate (say, for illustration, 30 years and 10% per annum respectively when A = 0.10608) or

(iii)   mills/kWh related to $/y by the assumed load factor (say 70% i.e. 6132 full power hours per year) and 1000 mills = $1.

For the relationship between these units see Table 3.

The units $/y per kW installed capacity will be used below. It will, where appropriate, e.g. for the uranium component, include the separate parts relating to the initial fuel loading, the replacement fuel and the last charge of fuel and the relevant lead and lag times.

Using the financial rules outlined above define the annual cost of electricity produced in an LWR in the once-through mode to be:

$$\{O\} = C + U + E + F + S + M \qquad (1)$$

where $\{O\}$ = total cost of electricity produced at a fixed load factor

C   = component due to capital cost of reactor including interest during construction, owners' costs, levy for decommissioning at end of life, etc

U   = uranium cost component of fuel cycle embracing initial, final and replacement quantities

E   = component of fuel cycle cost covering enrichment of the uranium

F   = fabrication cost component of fuel cycle for uranium oxide fuel

S   = storage charge and final disposal of irradiated fuel component of fuel cycle

M   = operation and maintenance cost.

Using the same financial ground rules, the cost of electricity produced in an LWR with reprocessing, recycling uranium and sending plutonium to storage for future use is:

$$\{R\} = C + U(1-x) + E(1-x') + F + R + W + M + H \qquad (2)$$

COST COMPONENT	$/kWe PRESENT WORTH	$/y PER INSTALLED kWe AT CAPACITY FACTOR Fe	MILLS/kWeh
$/kWe present worth	1	$\dfrac{1 - (1 + I)^{-L}}{I}$	$8.76 \cdot \text{Fe} \; \dfrac{1 - (1 + I)^{-L}}{I}$
$/y per installed kWe at capacity factor Fe	$\dfrac{I}{1 - (1 + I)^{-L}}$	1	$8.76 \cdot \text{Fe}$
mills/kWeh	$\dfrac{I}{8.76 \cdot \text{Fe}} \; \dfrac{1}{1 - (1 + I)^{-L}}$	$\dfrac{1}{8.76 \cdot \text{Fe}}$	1

Table 3.   Conversion factors for different generating cost representations

where x and x' represent the fractional savings in uranium and separative work due to recycling uranium

R = reprocessing and interim waste storage component of the fuel cycle

W = waste treatment and storage component of the fuel cycle

H = plutonium storage component of the fuel cycle (note: if plutonium is separated but not used for environmental or proliferation reasons its treatment and disposal would be included in W).

Using the same financial ground rules, the cost of electricity generated in an LWR recycling both uranium and plutonium is:

$$\{T\} = C + U(1-y) + E(1-y') + fP + (1-f)F + R(1+r) + W + M \quad (3)$$

where y and y' represent the fractional savings in uranium and separative work due to recycling both uranium and plutonium

P = fabrication cost component of fuel cycle for MOX fuel

f = the fraction of fuel to which P applies after making due allowances for the initial uranium charge

r = fractional change in reprocessing cost due to reprocessing MOX fuel and to any slight change in Pu recycle reactor fuel logistics.

Finally using the same financial ground rules, define the cost of electricity produced by a FBR to be:

$$\{B\} = C' + P_c + P_{rb} + R' + W' + M' \quad (4)$$

where C' = the capital cost of the breeder reactor, IDC, etc

P_c = core fuel fabrication component of fuel cycle

P_{rb} = radial breeder fuel component of fuel cycle

R' = average (core and blankets) fuel reprocessing and interim waste storage component of fuel cycle

W' = waste treatment and storage component of fuel cycle

M' = operation and maintenance component of fuel cycle

The fast reactor premium is now defined as the difference in the cost of electricity produced from the breeder and the cost of electricity produced from the once-through LWR where U is calculated at some arbitrary choice for the value of uranium - say \$25/lb U_3O_8.

If this premium is called Q, then:

$$Q = C' + P_c + P_{rb} + R' + W' + M' - C + U_{25} + E + F + S + M \quad (5)$$

Using the above definitions it can be seen that the boundary between the once-through and U recycle cases occurs at the uranium value where $\{R\} = \{0\}$, i.e.

$$U(1-x) + E(1-x') + R + W + H = U + E + S$$

i.e. $\quad U_{OR} = \dfrac{R + W + H - S - Ex'}{x} \quad (6)$

(Note $\quad U_{OR}$ = uranium component at the particular U_3O_8 value)

Similarly, the boundary between the U recycle and U and Pu recycle cases occurs at the uranium value where $\{R\} = \{T\}$, i.e.

$$U(1-x) + E(1-x') + F + R + H = U(1-y) + E(1-y') + fp$$

$$+ (1-f)F + R(1+r)$$

i.e. $U_{RT} = \dfrac{f(P-F) + rR - H - E(y'-x')}{y-x} \quad (7)$

and the boundary between the once-through and U and Pu recycle cases occurs at the uranium value where $\{T\} = \{0\}$, i.e.

$$U(1-y) + E(1-y') + fP + (1-f)F + R(1+r) + W = U + E + F + S$$

i.e. $U_{OT} = \dfrac{f(P-F) + R(1+r) + W - S - y'E}{y} \quad (8)$

Similarly it can be concluded that the boundary between the breeder and the once-through phase occurs along the line:

$$Q_{BO} = U - U_{25} \quad (9)$$

and the boundary between the breeder and reprocessing mode occurs along the line

$$Q_{BR} = U(1-x) - U_{25} - Ex' + R + W + H - S$$

i.e. $Q_{BR} = U - U_{25} - x(U - U_{OR}) \quad (10)$

and finally, the boundary between the breeder and thermal recycle phase occurs along the line:

$$Q_{BT} = U(1-y) - U_{25} - y'E + f(P-F) + R(1+r) + W - S$$

i.e. $Q_{BT} = U - U_{25} - y(U - U_{OT})$ \hfill (11)

It is known from simple algebra that, providing x and (y-x) are both positive, the quantity U_{OT} must, in numerical value, lie between the two quantities U_{OR} and U_{RT}, whatever the relative values of those two numbers are, because from equations (6), (7) and (8) the following algebraic relationship applies:

$$yU_{OT} = xU_{OR} + (y-x)U_{RT} \hfill (12)$$

Two possible cases are now distinguished.

Case 1

$$U_{RT} < U_{OR}$$

i.e. $\dfrac{f(P-F) + Rr - H - E(y'-x')}{y-x} < \dfrac{R + W + H - S - Ex'}{x}$ \hfill (13)

Alternatively, Case 2 where this inequality is reversed.

Case 2

$$U_{OR} < U_{RT}$$

i.e. $\dfrac{R + W + H - S - Ex'}{x} < \dfrac{f(P-F) + Rr - H - E(y'-x')}{y-x}$ \hfill (14)

Case 1 corresponds to the circumstances that probably apply in most countries. Roughly speaking, it corresponds to the assumption that the costs of reprocessing and waste management/disposal are more important factors than the subsequent cost of fabricating plutonium-bearing fuel.

Case 2 corresponds to the opposite assumption and under current economic conditions and using the expected range of fuel cycle component costs, it is unlikely that Case 2 will occur in practice. However, under certain circumstances, it is conceivable that reprocessing with uranium-only recycle could be economic. Such circumstances would entail high transport, storage and disposal costs for irradiated fuel in the once-through mode, low plutonium storage costs in the uranium-only recycle mode and high fabrication costs for MOX fuel assemblies (for instance, for the mixed island concept) in the plutonium recycle mode. A reduction in the interest/discount rate would favour this situation also.

    Phase diagrams corresponding to these two cases are shown
in Figures 15 and 16 respectively. In both these diagrams the
various lines are labelled according to the notation of the
equations just given. A glance at these diagrams shows that for
any chosen values of our main parameters, one of the four main
technology options will be preferred. The position of the boundaries between the various phases depends upon the parameters used
in the above equations and, of course, the values of those
parameters depend on technical or industrial judgements. However,
nothing we have done so far is dependent on future markets,
either for electricity or uranium ore. In practice, many discussions, particularly international discussions, are able to agree
on technical or industrial facts but are unable to agree upon
supply and demand considerations. The main value of this approach
is, therefore, that we are able to reach a broad consensus on the
appearance of the phase diagrams and this more clearly indentifies
where different countries are making different judgements because
of their views of the market situation.

    Nevertheless, it is stressed that within any particular
country the uncertainties in the economic data and the variations
in reactor parameters would change the sharp lines into broad
band-widths but, for simplicity of presentation, only the lines
are shown in these diagrams. Furthermore, it should be stressed
that technical judgements vary substantially from one country to
another and the economic ground rules vary sufficiently from one
country to another so that, quite properly, different countries
will show the boundaries between the various thermal reactor
phases at somewhat different positions.

    So far in all the above equations and phase diagrams, the
value of plutonium has been assumed to be zero. We now must
consider how the analysis is changed when plutonium has become
a valuable commodity, i.e. when fast reactors have been launched
or when thermal recycle of plutonium is being practised.
Clearly, there is no change in the cost of electricity produced
by the once-through mode because, in this mode, the plutonium
is not separated from spent fuel. However, a nation or utility
which sends fuel for reprocessing may decide to sell its plutonium after storing it for a while and, in this case, equation
(2) becomes modified to the following:

$$\{R'\} = C + U(1-x) + E(1-x') + F + R + W + M + H - V \qquad (2')$$

where V is the value received for the plutonium produced per year
per kW installed capacity.

    The value of plutonium in the case of thermal recycle is

already implied by the use of equation (3) but equation (4) needs
an important modification. In addition to the capital cost needed
initially to build a fast reactor, we also need to take account
of the initial cost of acquiring the plutonium needed to launch
it. Suppose that the initial plutonium inventory of the fast
reactor requires the operation of a thermal reactor for t years
then we have the additional launching cost of tV and an additional
term tVI has to be added to equation (4). However, that is not
the only modification which needs to be made because we also
need to note that if the fast reactor is breeding then we need
give it credit for the excess plutonium which is produced. This
credit produces an additional term in (4) which is tV/τ, where
τ is the doubling time. In total, therefore, in place of (4) we
get

$$\{B'\} = C' + P_c + P_{rb} + R' + W' + M' + tV\{I\frac{1}{\tau}\} \tag{4'}$$

We now know that V takes on different values in different
parts of the phase diagram. Along the boundary between the once-
through and breeder phases, the appropriate value of V to be used
in (4') is the cost of obtaining the plutonium. This is simply:

$$V = \{R\} - \{O\} = x(U_{OR} - U) \tag{15}$$

Along the boundary between the reprocessing mode and the fast
reactor mode, we have the formula:

$$V = O \tag{16}$$

and along the boundary between thermal and fast reactors the value
of V which must be used is the opportunity cost of using the plutonium to launch fast reactors instead of using it in thermal
recycle. Hence we get the formula:

$$V = R - T = (y-x)(U-U_{RT}) \tag{17}$$

Inserting these three values into equation (4') modifies the
phase diagrams as shown in Figures 17 and 18 and, in the remainder
of this discussion, we shall refer to those two figures.

So far we have determined the value of plutonium only along
the boundary lines between the various phases. We can extend
the analysis to give a value for plutonium inside any particular
area of the phase diagram only if we make some assumptions of the
terms under which plutonium will be traded. For the sake of
simplicity, let us assume that it is traded at a free market value.
Then a utility can gain value from its plutonium either by using
it itself or by selling it to other utilities. We can therefore
determine the value of V by simply equating $\{R'\}$ either to $\{O\}$,

{R}, {T}, or {B'}. This tells us that the cost of obtaining
a unit of plutonium (either real cost or lost opportunity cost)
is either (15), (16) or (17) in the {O}, {R} and {T} phases
respectively. Within the breeder phase, equating (2') to (4')
gives us the simple result:

$$V = \frac{\{R\} - \{B\}}{1 + tI - t/\tau} \qquad (18)$$

DISCUSSION

Let us now use this algebraic background to represent the
various views which have been expressed by various nations about
the cost of fast breeder reactors and their associated fuel cycle
services, the timescale by which they might be available and the
simultaneous variations in the price of uranium. Various views
expressed by groups in different countries can be superimposed on
the phase diagram and three illustrative trajectories marked X, Y,
and Z are shown on Figures 19 and 20.

In these trajectories, time increases along the line and in
general uranium prices move upwards and the fast reactor premium
moves downwards as the benefits of R & D, replication and economies
of scale take effect.

The line marked X illustrates the view of a country which
expects neither a rapid reduction in fast reactor generating costs
nor a rapid increase in the price of uranium, so that the once-
through cycle or thermal recycle will be the cheapest option for
many years ahead.

The line marked Y illustrates the view of a country which
expects somewhat lower fast reactor generating costs, but a
substantial increase in the price of uranium, such that uranium
or uranium and plutonium recycle become the most attractive
option.

The line marked Z illustrates the view of a country which has
developed fast reactor technology early and expects fast reactor
generating costs to decrease rapidly with time accompanied by an
increase in the price of uranium (particularly in a world with few
or no fast reactors) so that the fast reactor fuel cycle will
become the most economic choice at an early date.

Actual presentations made by individual countries in recent
discussions serve to verify the general nature of this discussion
and a detailed examination of the various views which have been
expressed brings out two points very strongly. First, if nuclear
power is to make a long term contribution to the needs of the

world, the introduction of fast breeder reactors is both essential
and necessary. At first sight, views may appear to differ about
the economic viability of fast reactors but in the end those
disagreements merely result in slightly different assessments of
when they will be important on a large scale. Some countries
judge that this time will be as early as 1995 and others judge
it will be postponed until 2025. A difference of only 30 years
is trivial measured against the lifetime of mankind. This
simple thought has been expressed in various ways many times and
is easily seen from the phase diagrams. If we move far enough
to the right, the breeder reactor must be favoured. Most published
discussions about the economics of fast reactors spend a great
deal of time on this point and necessarily involve uncertain
judgements about the variation of uranium ore price with time.

However, trajectories shown in Figures 19 and 20 demonstrate
another factor of equal or more importance. It is that, as the
scale of fast reactor use increases, the cost of electricity
produced by fast reactors will decrease dramatically. Therefore,
an essential feature of the nuclear strategy of any country
must be its industrial strategy for decreasing the capital cost
of fast reactors and fast reactor fuel cycle centres. It is,
therefore, very important with fast reactors to make the correct
judgement of when to introduce large scale industrialisation.
This in turn means the economics of fast reactors will be very
significant to individual countries and very dependent upon their
industrial capability and their manufacturing expertise.

Generally speaking, in many discussions concerning fast
reactors this last point is not given the weight it deserves.
There is nothing very surprising about this. It is freely acknow-
ledged over a wide variety of endeavour that people doing research
and development to introduce new technologies always underestimate
the difficulty of converting their technology to a routine production basis and they underestimate the difficulty of penetrating
the market place with that new technology. The formalism of
these phase diagrams simply obliges us to acknowledge those points
specifically and give some estimate of them.

FAST BREEDER REACTORS - LECTURE 5

W. Marshall[*] and L.M. Davies[+]

UKAEA[*]
11 Charles II Street
London, SWIY 4QP, U.K.

AERE Harwell, Didcot[+]

INTRODUCTION

We now turn to the most difficult lecture of this series concerning fast breeder reactors because we are now going to discuss questions which are not scientific, not technical nor even economic; we are going to discuss a range of issues which, in recent years, have alarmed some members of the general public about the use of fast reactors. In short, we shall consider the emotional question of the "plutonium economy". Opponents to fast reactors do not always make their points clear but so far as we are able to understand them, the "social" objections to the fast reactor fall into one or all of six areas.

  (i)   Fast reactors are too expensive.

 (ii)   Fast reactors produce nuclear waste which in some way, it is suggested, will be worse than waste produced by thermal reactors.

(iii)   The use of plutonium introduces "the plutonium economy" and that opens the possibility of terrorist groups stealing the plutonium and using it for blackmail purposes.

 (iv)   The difficult security measures needed to protect the plutonium from being stolen will themselves limit our civil liberties in some fundamental way.

  (v)   The use of plutonium in a worldwide way will assist the proliferation of nuclear weapons.

(vi)   Fast reactors are such sophisticated machines that they
        cannot provide energy for developing countries because
        the latter would not have the expertise either to
        operate or build them.

We have discussed the question of fast reactor economics in
the previous lecture. The other points are not technical and
many of our scientific colleagues tend to dismiss them as insub-
stantial or irrelevant. However, that in our opinion is not a
satisfactory way of dealing with legitimate concerns and, therefore, although these points are neither quantitative nor easy to
discuss, we shall now do our best to discuss each of these in turn.

NUCLEAR WASTE

We have lost count of the number of times when, in general
conversation with members of the public in the UK, they have
commented that the fast reactor produces a more difficult waste
problem that a thermal reactor. We do not know where this view
has come from and neither do we know if it is specific to the UK,
but nevertheless, let us first address ourselves to this question.

Both thermal and fast reactors produce heat and thereby
produce electricity by the controlled use of the fission process.
In the case of thermal reactors, the main fission process is
that involving ^{235}U, whereas in fast reactors the main fission
process involves ^{239}Pu. The fission process automatically
produces fission products, i.e. two nuclei each roughly half the
mass of a uranium or plutonium nucleus. Those fission products are
highly radioactive and produce the most dangerous part of the
nuclear waste from either type of reactor. However, the nature of
the fission process is such that it cannot significantly differ
between uranium and plutonium and there can, therefore, be no
significant differences between the fission product waste produced
be either type of reactor. Any slight differences that do exist
are second-order effects which can be ignored at this level of
discussion and in practice. There is, therefore, no reason in
principle or in practice why fission product waste should differ
significantly from one case to the other.

However, the fission process is not the only one we need to
consider. In addition to that, nuclei can simply absorb a neutron
to produce what are called transuranium elements, i.e. nuclei which
weigh more than uranium. The most important transuranium element
to be produced is simply ^{239}Pu. This is the plutonium isotope we
have discussed throughout this lecture course and its production is
inevitable because all nuclear fuel contains ^{238}U, all ^{238}U absorbs
neutrons and that neutron absorption process converts ^{238}U to ^{239}Pu.

    In addition to that, ^{239}Pu can itself absorb a neutron to
become ^{240}Pu and that in turn can absorb a further neutron to become
^{241}Pu which becomes ^{241}Am, etc. The production of these actinides
can vary from one reactor to another because the neutron absorption process depends upon the energy of the neutrons which are
present and the successive incineration of these actinides also
depends upon the way in which the reactor is operated. For example
if fuel is put into a reactor and taken out of it after a
relatively short period then a certain amount of ^{238}U will have
been converted to ^{239}Pu but there would have been no time for that
^{239}Pu itself to be converted to ^{240}Pu. The production of actinides
can, therefore, vary between fast reactors and thermal reactors
but, for all practical purposes, this production of actinides
always starts with the simple conversion of ^{239}Pu. In all cases,
therefore, the dominant actinide which is produced is ^{239}Pu. We
have seen in the earlier lectures that that is produced in
thermal reactors and can be incinerated efficiently in fast reactors.
If that ^{239}Pu is not incinerated then it is a waste product. In
other words, if there are no fast reactors and no incineration of
^{239}Pu, then the quantity must increase indefinitely. In short,
therefore, if we do not have fast reactors then the plutonium
produced by a thermal reactor must be classified as a waste
product. By that judgement, it is clear that fast reactors must
produce less waste products than thermal reactors alone. It
follows from this that there is no corresponding ecological
argument which can be produced against the fast reactor as compared
to thermal reactors. Indeed, the argument we have just given
underplays the advantage of the fast reactor by a large margin
because, on a worldwide basis, the single largest detriment to the
environment produced by nuclear power is almost certainly the
release of radioactivity from the mining of uranium. Since the
introduction of the fast reactor avoids new mining operations as
far as possible, it follows therefore that the introduction of fast
reactors has actually a positive effect on the ecology of the
world relative to the use of thermal reactors.

TERRORISTS AND PLUTONIUM

    This concerns the fear that plutonium will be stolen or
hijacked by terrorists or sub-national groups which would hold
the civilised world to ransom. It is unfortunately true that we
must give attention to that possibility and it is sad to reflect
that the growth of violence throughout the world has brought us
to that acknowledgement. It is, however, not the only place
where we are obliged to adopt measures which are intrinsically
distasteful. Air travel nowadays is made unpleasant by the
necessity for luggage and body searches because of these terrorists
and sub-national groups. Nowadays national leaders in almost all

countries have to be guarded night and day from attempts at
assassination or kidnapping. We find this one of the most
miserable features of the modern world. It does, however, have
more to do with the existence of terrorists than the existence of
plutonium and although this distasteful subject must be treated
with the seriousness it deserves, it does not seem to us that the
existence of this danger should determine the whole future course
of civilisation. Furthermore, the danger is, in our opinion,
exaggerated. The GESMO study in the United States concluded
that if the use of plutonium was introduced on a large scale in
the United States then it could be guarded to a satisfactory
level without undue difficulty. This is a conclusion which is
likely to be generally valid. It does give special difficulties
for a country like our own where we have a long standing tradition
that for normal civil activities we do not arm our police. We
have no doubt that this well known tradition encourages criminals
in the UK not to carry firearms, but most international terrorist
groups intent on stealing plutonium would almost certainly use
firearms. It follows, therefore, that the steps taken to ensure
physical protection of significant quantities of plutonium,
whether in store or in transit must include not only physical
isolation but also suitably armed policemen. This is an uncomfortable conclusion to come to in the UK, but in other countries
of the world, where police are normally armed, it does not raise
new questions of principle. Therefore, in our judgement and in
the judgement of many responsible people who have studied this
matter worldwide, this is not an adequate reason for avoiding
the use of fast reactors.

The necessity to guard plutonium which we have just discussed
should not, however, mislead people to the idea that the contruc-
tion of a nuclear weapon using plutonium is an easy matter. It
is a misconception to suppose that a small number of people who
obtain plutonium could convert it into a weapon. The task of
doing so is actually very complicated and difficult. We have
seen a number of ideas produced by "amateur bomb designers" and,
in our judgement, none of them would explode. For obvious reasons
we do not want to discuss this subject in public but it is worth
making a few comments to put the problem in perspective. These
comments are all based on information which is freely available
in the open literature.

If we cast our minds back to 1945 when the Manhattan Project
had been operating for a few years and had prepared its first
few weapons, then we know that they had prepared two types: one
based on ^{235}U and one based on military grade plutonium. The
project had assembled the very best scientific brains and engineering
expertise at Los Alamos and made two decisions. They were
entirely satisfied that the uranium weapon would explode and
there was no need to test it but they were not confident that

the plutonium weapon would explode and they therefore felt that a test of that was essential. The test in New Mexico desert was actually successful and, therefore, one uranium weapon and one plutonium weapon were used at Hiroshima and Nagasaki respectively. However, the point of the story lies in the fact that, all that assembled expertise of the most brilliant scientists in the USA, Canada and the UK thought that the design of the plutonium weapon was sufficiently difficult that they were not confident in advance that it would detonate. It is fair to deduce from this that the characteristics of a plutonium weapon, which necessarily must remain classified, are more difficult and more complex than that of a uranium weapon. It is our judgement, therefore, that if a group of terrorists attempted to make a crude weapon from plutonium then, almost certainly, all they would do is produce a "fizzle". It is for this reason that we believe the risks coming from plutonium diversion are generally exaggerated in the popular literature.

CIVIL LIBERTIES

It has been argued in our own country and to some extent in others that the security measures which are required to protect plutonium will lead to an abuse of civil liberties in some fundamental way.

This concern is best answered by the reply given by Mr. Benn (the then Secretary of State for Energy) in 1977 to such questions and is quoted below:

"As will be seen from the replies given to earlier questions, it is not considered that the development of a large nuclear programme would give rise to security arrangements materially different from those at present. The level of the national security commitment would not however be dictated only by the size of the nuclear programme or the extent to which it employed plutonium, but by the scale, the nature and the intensity of terrorism and other violent criminal activity. It cannot be assumed that such terrorist activity would necessarily be directed against nuclear plant and material, or that, if the latter did not exist, terrorism and other violent crimes would disappear. And although the general situation in this country and in the world in the first half of the next century is not of course one which can be predicted with exactness and confidence, no democratic Government would want to take precautions against terrorism and violent crime that would be more derogatory of individual liberties than the actual situation demanded."

PROLIFERATION

Proliferation involves the acquisition of nuclear weapons by governments and the possibility of this is a real and geniune question which must be discussed carefully and responsibly. Because fast reactors produce and use plutonium, it follows that they could be abused to produce nuclear weapons. However, it does not follow that the problem would be better if we abolished the use of fast reactors. Let us now discuss this matter as carefully as we can by reviewing all the arguments that have been produced in recent years on this issue, trying to put them in perspective one to another.

The first point to make is that a decision by a government to construct nuclear weapons is a political decision motivated by political considerations. Whether a government has nuclear materials immediately available to it or not is almost irrelevant. The connection between civil nuclear power and nuclear weapons proliferation must, therefore, necessarily be a marginal matter. It is nevertheless sufficiently important that we cannot dismiss it.

It should be remembered that a dedicated nuclear weapons programme is cheaper and quicker than an approach to nuclear weapons which first sets up civil nuclear power facilities. No country with weapons ambitions would, therefore, normally choose to start off with a civil programme. Nevertheless, the existence of a civil nuclear programme has, by its very nature, produced training know-how and skills which could be diverted to a weapons programme. However, it seems impossible to argue for that reason alone that civil nuclear power should be discouraged. One could just as well argue that any form of higher education of a technical kind could be abused by a government into a weapons programme. Yet we do not argue that the use of universities should be abolished throughout the world because of that reason. There is, therefore, no definitive argument of this kind which can be made.

Next we should point out that the diversion of plutonium is not the only way in which nuclear weapons could be constructed. Easier enrichment techniques for ^{235}U are becoming more widely appreciated and we should remember that the difficult step in isotope enrichment of uranium is to do it on a large scale at a suitably cheap cost. To do it on a small scale is nowhere near as difficult and many people who have thought about this subject believe that the use of enrichment techniques to separate ^{235}U presents a greater proliferation risk than the use of plutonium. There is indeed a real danger that, by concentrating on the plutonium issue, other routes towards nuclear weapons will not be inhibited as thoroughly as they should be.

Turning, however, to the diversion of plutonium to produce
weapons, we should first recognise that the production of plutonium
is an inevitable consequence of nuclear power. As soon as uranium
of any isotopic composition is put into a reactor, ^{238}U is irradiated and creates ^{239}Pu, indeed about half of all nuclear power
comes from plutonium fissions. Many arguments which have been
orientated against the fast reactor on these grounds are, in our
opinion, misdirected. The plutonium is created in the first place
in thermal reactors and if we are to have nuclear power at all,
we are obliged to accept that that is a fact.

Some people have argued that the once-through cycle in which
spent thermal reactor fuel is not reprocessed is a desirable
technology because the plutonium does not exist in separated
chemical form but remains within the spent fuel element. It is
therefore intimately mixed with fission products of high radioactivity, which act as a self-protecting "policeman" to prevent
plutonium diversion. In our judgement, however, this argument
does not survive careful examination. It is not possible for
spent fuel elements to be highly radioactive indefinitely.
Indeed the radioactivity of a PWR spent fuel element falls off
as shown in Figure 21 and, after a suitable interval of time, the
fission product activity is not an automatic protection for the
plutonium. Eventually, therefore, the plutonium produced in a
thermal reactor becomes "accessible" and, from a proliferation

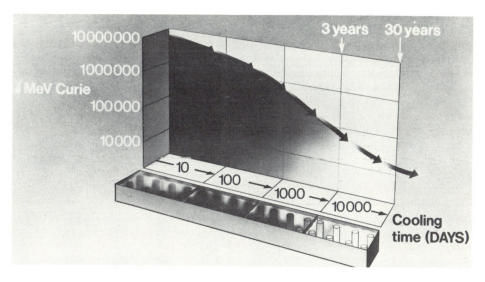

Fig. 21.   Gamma activity of a spent PWR fuel element with time
            after removal from reactor

point of view, we do not believe it is wise to advocate the
indefinite and widespread use of the once-through cycle.

Furthermore, we should recognise that the once-through cycle
produces more plutonium in the world than either thermal recycle
or fast reactors. The reason for this is self-evident. In either
thermal recycle or in fast reactors the intent is that plutonium
is incinerated to produce heat and then electricity. Going back
to Lecture 1, you will recall that the production of plutonium
from a fast reactor was in the range of 170-250 kgs per GW year.
The corresponding figure for light water reactors is 330 kgs per
GW year. The introduction of fast reactors, therefore, reduces
the amount of plutonium in the world. It is remarkable how many
times the opposite statement is made in public discussions.

However, of course, it is true that a country operating fast
reactors is handling plutonium or perhaps mixed uranium/plutonium
oxide in a readily accessible form and it is true that that is
easier to convert to a weapon than the plutonium contained in a
spent fuel element. However, we have already remarked that the
radioactivity of a spent fuel element decays with time and,
therefore, the distinction between plutonium in these two forms
becomes less sharp as time progresses. Furthermore, a nation
which has the ability to build, or even just operate, fast reactors
is a nation of a high technological standard. For a nation of that
standard the form of plutonium, whether it is accompanied by
radioactivity or not, is a trivial and unimportant question. In
contrast to that, if the government of any particular state
found the form of "accessibility" of the plutonium to be a suitable
inhibition to a weapons programme, then that state would not
anyway be at an advanced enough level to have a fast reactor in
the first place and, therefore, the question does not arise.
This last point illustrates the point that many published arguments
concerning the proliferation questions are highly theoretical and
academic in their approach. We should not actually be comparing
proliferation risks of fast reactors in use now with thermal
reactors in use now because it is a hypothetical comparison. Fast
reactors will not be in use on a wide scale until well into the
next century. We should, therefore, be comparing and discussing
the proliferation risks of the fast reactors as used then and
the once-through cycle as used then and, at that point in time,
an "accessible" form of plutonium will be that contained in old
spent thermal reactor fuel elements.

To sum all this up, we surely must conclude that any nation
which could successfully develop and introduce a nuclear power
programme has the ability to undertake successfully the much
easier task of having a nuclear weapons programme. We return,
therefore, to the first point we made. The proliferation of
nuclear weapons is primarily a political issue and the true

answer to any proliferation question is the motivation and the
determination of any individual national government and the
consensus view amongst governments that the widespread proliferation of nuclear weapons should be avoided.

FAST REACTORS AND DEVELOPING COUNTRIES

     This is the last item we shall discuss in these lectures.
In the first lecture, we stressed the ability of fast reactors to
meet the energy demands of the world community. However, the
later lectures have probably satisfied you that fast reactors are
quite difficult to build and are complex machines which depend not
only on high quality fabrication of the reactor and highly trained
operators to run them, but also require an elaborate fuel cycle
to sustain them. It is, therefore sometimes argued that the
development of fast reactors is totally irrelevant to the needs
of developing countries. In our own country, some people go on
to argue that, for this reason, research and development activities
effort should be put into other simpler technologies. In our
opinion, this argument does not stand up to examination. Let us
examine why this is the case.

     In the previous lecture we discussed the economics of fast
reactors using phase diagrams and we described various industrial
strategies which might be adopted by advanced countries to
introduce them. Suppose now that the most advanced countries
take policy decisions to introduce fast reactors then, as each
advanced country switches to fast reactors, it becomes removed
from the world uranium market and the normal laws of supply and
demand will then lead to a stabilisation of uranium ore prices.
That immediately produces a benefit to other countries, presumably
developing countries, which have acquired enough expertise to
operate thermal reactors, but cannot yet contemplate introducing
fast reactors. They are then able to purchase their uranium ore
at a more reasonable cost than would otherwise be the case.
(This is not such a good point for those developing countries
which also happen to be uranium exporters, but they are a
special case).

     However, this is not the only benefit produced for the
developing country. Going back to the results of the first
lecture, we must remember that the plutonium inventory to launch
a fast reactor is large and the rate of breeding additional
plutonium is actually slow. Therefore, for a wide range of
figures for the economic growth rate of the advanced countries,
the introduction of fast reactors is limited by the availability
of plutonium. In that situation, plutonium acquires an international market value for the production of electricity. This
in turn means that spent thermal reactor fuel acquires a market

value. We therefore, have reached a situation where the advanced
countries who want to build more fast reactors need plutonium to
do so whereas developing countries who have thermal reactors
are producing plutonium they have no use for. The obvious thing
to do in those circumstances is trade in plutonium and the simple
diagram of Figure 22 shows that, in a free market model, the
benefits of that plutonium trade fall to both the advanced and
the developing countries. Indeed the simple algebra of the
previous lecture and illustrated in diagrammatic form in
Figure 22 shows that the developing country has the full benefit
of fast reactors without having fast reactors at all! In this way
we can conclude that fast reactors do in fact benefit developing
countries provided we can, in practice, ensure that this trade
in plutonium comes about in the way outlined by this discussion.
Looking back to equation (18) of the previous lecture, we see
that the essential conditions for these circumstances to arise
are the following:

(i)   The cost of producing electricity from the breeder {B}
      must be less than {R}. This can be the case provided
      the cost of uranium has drifted up far enough and the
      benefits of scale have been fully effective in launching
      the fast breeder programme.

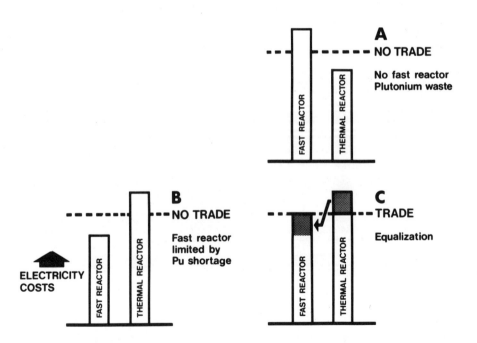

Fig. 22.   Equalisation of electricity costs by world trade in
            plutonium

(ii)   The economy of the advanced country must be expanding so
        that it needs more electricity. If this were not the case,
        it would not need to build more fast reactors and, there-
        fore, the incentive for trade would disappear.

(iii)   The doubling time must not get too short. On present
        design of fast reactors using oxide fuel, this does not
        seem much of a risk. Doubling times are frequently
        estimated to be in the 20-40 year range and a long
        doubling time obliges advanced countries to rely upon
        a source of plutonium from thermal reactors to expand
        its electrical generation capacity.

However, as we mentioned earlier, it is possible to design
fast reactors with short doubling times. This certainly would be
the case if we introduced carbide fuels. If that were done too
early, there is a risk that it would produce a glut of plutonium
and the argument we have just reviewed would collapse.

It follows from this argument that the intelligent use of
fast reactors can benefit both advanced and developing countries.
However, this can only be the case if they are introduced by the
advanced countries in such a way that they encourage inter-
dependence between nations of the world. There is, therefore, no
guarantee that fast reactors will help developing countries and
this discussion has simply served to point out that the opportunity
exists. Whether that opportunity is taken or not will depend
upon governments and not upon scientists.

ACKNOWLEDGEMENTS

We are indebted to Mr A M Judd and Mr A A Farmer, together
with many others in the UKAEA, for their contributions and comments
which have helped in the preparation of these lecture notes.

HOW REACTOR SAFETY IS ASSURED IN THE UNITED STATES

Robert J. Budnitz
U.S. Nuclear Regulatory Commission
Washington, DC 20555

This is the first of four lectures* that I will give in this School. The main goal of this talk will be a description of the present system in the United States for assuring the safety of large nuclear power reactors. The second talk will cover risk assessment and its use in reactor safety and regulation. The last two talks will discuss the recent accident at Three Mile Island. This talk will begin with a description of the underlying safety philosophy, including which institutions have the various responsibilities. I will then proceed to discuss how the safety assurance system is implemented, how oversight is accomplished outside of the system itself, and how feedback occurs to improve the system. Finally, some issues and problems with the system will be discussed.

I shall not be covering the broader question of whether nuclear electric power plants are an appropriate, or best, choice for generating electric power in our country or any other country. I feel that this basic policy issue requires much more input than that which I have chosen to discuss here today. However, some of the points I shall be making can be of guidance to any country that chooses to undertake a nuclear power program, or that has already made that decision. This is because many of the institutional issues raised have broad generic implications that extend beyond the borders of the United States, with its own particular political arrangement.

Also, I must emphasize at the outset that the opinions expressed here are my own and do not necessarily represent policies of the U.S. Nuclear Regulatory Commission. I shall attempt to be candid with you concerning both the strengths and weaknesses of our present safety regulatory system, as I see them myself.

_____
*Only the first two lectures are included in this volume.

The talk will focus on the American system mainly because that is the system with which I am most familiar. Other countries' safety and regulatory systems will not be treated, although many important lessons useful in other countries can be easily transferred from the American experience. Also, because the large commercial reactors in the United States are mainly PWRs (pressurized water reactors) and BWRs (boiling water reactors), these will be the focus of attention. The remarks may not be fully applicable to the gas reactors prevelant in the United Kingdom, to the heavy water reactors built by our Canadian neighbors, or to the fast metal-cooled reactors now under intensive development in many countries.

The emphasis will be on safety through regulation. I will thus give less attention to issues as they might be presented by a reactor vendor, an architect/engineering firm, or a utility/owner. I hope that this admittedly biased approach will nevertheless be of interest to the attendees at this School.

The main design concept under which power reactors have been built and operated in the U.S. is known as the "defense-in depth" concept. This concept has three elements. The first is the philosophy that the designer should contemplate all important accidents that he thinks have significant probability of occurring, and should design against them. The designer should use intrinsically safe designs wherever possible, and engineered systems elsewhere. The concept of an intrinsically safe design implies reliance where feasible on fundamental physical laws for protection. A good example of this is the fact that the water in the reactor coolant system also serves as the moderator in U.S. power reactors, and its loss automatically shuts down the chain reaction. This is an intrinsic feature that does not require specific engineered equipment to carry out its function. Of course, not all elements of the design can rely on such physical phenomena for protection against accidents.

The second element in "defense-in-depth" is the assumption that, despite best efforts in design, accidents will happen anyway, because they cannot be all designed against absolutely (either intrinsically or through engineered features), and therefore significant redundancy must be built into the designs to cope with equipment failures. Also for many important types of accidents there is a need for engineered safety systems to cope with events that occur. Examples of these systems include the well-known ECCS (emergency core cooling system), the auxiliary feedwater system, seismic restraint systems, and redundant/diverse systems to terminate the chain reaction.

The third element in "defense-in-depth" is reactor design and siting to mitigate consequences of accidents if they occur despite the best efforts of the first two approaches. Included in this part of the philosophy are the large reactor containment building, built to contain radioactive releases if they occur from the reactor itself

the use of sites remote from large populations, so as to reduce the
doses that potentially might be delivered to nearby residents in the
event of a radioactive release; and the filtering of airborne releases to reduce their magnitude.

One of the key design features of our reactors is the multiple
barriers against the release of radioactive materials from the fuel.
There are four barriers between the fission products in the fuel
and the broader environment. The first is the fuel matrix itself,
which is a ceramic-like material that immobolizes or significantly
retards the various fission products. The second is the fuel cladding, designed to retain within itself the fission products that
do migrate from the fuel matrix. The third is the reactor primary
pressure vessel system, designed to keep fission products which might
escape from the fuel cladding from going outside the reactor; and
the fourth is the containment building mentioned earlier. Each of
these barriers will contain or retard radioactive releases, and their
effectiveness is a major element in safe reactor design and operation.

There are three crucial considerations in putting together a
safe design for a reactor, and careful attention must be paid to
each. The first is the ability to shut down the chain reaction and
keep it shut down; this "scram" mechanism must be reliable and effective. The second is maintaining structural integrity of the fuel, the
core structure, the primary coolant boundary of vessel and piping,
and the larger building and its equipment. Especially when one considers seismic or hurricane forces, but also in considering forces
generated internally to the reactor, this structural design problem
is a very difficult one. The third is the absolutely crucial need
for residual heat removal after reactor scram. Unlike fossil-fired
power plants whose fuel can be fully extinguished, nuclear plants
cannot fully extinguish the heat generated in the chain reaction;
there remains a very large body of fission products within the fuel
after scram, and these produce large amounts of heat as they themselves undergo radioactive decay. This "decay heat," amounting to
several percent of the initial heat energy just after scram, and as
much as 2 percent a day later, must be removed or the entire reactor
core will ultimately (indeed, rapidly) rise in temperature to the
point where fuel damage will release the encapsulated radioactivity.
This need for maintaining a long-range heat sink is recognized as one
of the most vital issues in recovery from accidents large or small.

Finally, there is the issue of reliance on automatic features
and its complement, reliance on the human operator. Reactor designs recognize that many safety-related functions are best made automatic. Examples include the automatic actuation of the several ECCS
systems when various pressure set-points are crossed, and automatic
actuation of auxiliary feedwater pumps when the main feedwater pumps
fail. Despite using such automatic systems, reactor designers

recognize that there are numerous situations (indeed, much more numerous than those for which automatic systems can be designed) in which human operator intervention is essentially <u>required</u> for safe operation. This is because the complexity of possible accident sequences is far too great to be contemplated by the designers, and human judgment with its versatility and integrating powers produces safer operation. One key issue, demonstrated at Three Mile Island, is that the trade-offs between automatic and human system operation always involve compromises, and sometimes the present arrangement leaves much to be desired.

Integral to the American reactor program is a strong reliance on quality control and quality assurance, which begins with the design phase and extends to manufacture, construction/installation, operations, and maintenance. This program, carried out by the vendors, constructors, and utilities, is monitored in an audit mode by the regulatory groups. It has resulted in components and systems in our nuclear plants that are significantly more reliable and better performing than comparable systems in related nonnuclear industries. In a real sense the quality programs are a backbone of the whole endeavor.

Given this outline of the fundamental philosophy of reactor design and operation, it might sound as if accidents could almost never happen. Indeed, it might seem as if the system should be capable of nearly perfect operation, excepting of course minor breakdowns of components and systems that don't compromise safety very much. This conclusion would not be correct. Even with the most hurculean efforts described above, our reactors are very far from perfect machines. We have continued to find errors of various types in all phases of the program: conceptual errors; improperly used engineering assumptions; errors in engineering computer codes; errors of manufacture, construction, or installation; maintenance errors; operator errors; and so on. Brief reflection reveals that these are to be expected. Indeed, if one somehow stopped finding these errors it would be a sign of the failure of the error surveillance system! The key task of safety assurance, then, is in three parts: to assure that the system is <u>forgiving</u> of and can <u>recover</u> successfully from those errors that do occur; to assure that when errors not forgiven appear that they are <u>corrected</u>; and to assure that continuing <u>improvements</u> in all elements of the system are developed and implemented.

Much of what I will have to say next about the success of our reactor safety program will be praise for the success with which these three elements (forgiveness, correction, improvement) are carried out. Much of my criticism of the present safety arrangements is, by contrast, criticism of these same elements.

In the U.S. the fundamental responsibility for safe construction, maintenance and operation of the reactor rests with its owner,
the utility. This responsibility is shared in a very complex way
with other elements of the nuclear industry: the reactor vendor,
the architect/engineer, the industry organizations, and so on. The
sharing is complex because even though the utility may have only
limited expertise in some highly technical areas, and may therefore
rely on other groups (some under contract, some not), it bears the
major legal responsibility. It is important to point out that we
have in the U.S. a couple of hundred electric utilities of significant size, many of which are now owners of reactors either operating or under construction. The utility industry is fractionated
into the large number of individual companies because of the way
the industry grew up in its early years. Some of the utilities
are privately owned corporations; some are publicly owned; and a
few are Federal agencies. Some utilities are large and very strong
technically, while others are small and relatively weak. This creates its own problems, since the success of the entire reactor endeavor in the U.S. depends on achieving comptence even with the
smaller, less well-financed utilities.

Another element of diversity is the variations among our commercial reactors themselves. We have four manufacturers of light-water
reactors, each with several different designs; and we have about a
dozen different architect/engineering firms that put up the plants,
all with different construction practices and varying designs for
the nonnuclear parts of the facilities. Thus, nearly all of our
reactors (about 70 operating, another 125 under construction or
planned) are different. The fact that hardly any are alike seriously
affects the whole U.S. nuclear safety picture. For example, analysis of safety performed at one reactor often cannot be directly
applied to more than a few others, and problems overcome at one
reactor may not be applicable at any other plant!

The federal government, through the U.S. Nuclear Regulatory
Commission, has overwhelmingly preeminent authority in nuclear
regulatory matters. Although we have a federal system in the U.S.,
the various states have little impact on safety, with the single
(but crucial) exception of their key roles in site selection, environmental regulation, and in rate-setting, which are done in whole
or in part at the state level. This federal preemption of nuclear
safety regulation is mandated by act of Congress.

The Nuclear Regulatory Commission (NRC), where I work, has a
general philosophy of regulation that is best characterized by the
word "audit." That is, although it establishes large numbers of
rules, regulations, guides, and standards, the method it uses to gain
assurance that the reactors are properly run is by doing audit reviews of various elements of reactor design, construction, and
operation. These reviews, not intended to be completely thorough

enough to provide full assurance of all details, are intended to assure the NRC that the utility and its contractors do _their_ job properly.

Unfortunately, the system which gives the utility the prime safety role really doesn't work the way it is supposed to work. What has happened over the years is that the NRC (and its predecessor agency) have imposed a complicated set of regulatory conditions on the utilities.

Many of the utilities consider that the amount of regulation is much too great, that it constitutes an unreasonable burden which does not contribute to safety in proportion to its size and cost, and that in many instances it represents overkill. This attitude, quite common before the accident at Three Mile Island, is still present, although less dominant, even after TMI. Its effect has been to place the regulatory program in the driver's seat insofar as safety is concerned: the utilities seldom have taken initiatives beyond what is required by the NRC, and the NRC spends inordinate amounts of effort worrying about the imposition of requirements that might, in another arrangement, have been undertaken by the utilities on their own initiative. This situation has changed somewhat since Three Mile Island, of course, but whether it will be a permanent change remains to be seen.

The NRC is not required by its philosophy (or the enabling legislation) to regulate so as to make reactors as safe as they can reasonably be; it is only required to make findings that the safety of a particular system provides "no undue risk to" the public health, safety, and the environment. This latter requires judgment, and here again one meets a conflict: some members of the industry and the public believe that the safety of our large commercial reactors and is already far more than "safe enough" to protect the public and much resentment of additional regulatory requirements exists. On the other side of the coin, the regulatory staff is accused by some concerned groups and citizens of being "soft" on the industry, because of its close working relationship with them. Often cited, for example, is the NRC's reluctance to order back-fit of safety improvements. This tension and its effects are a major part of U.S. reactor safety and its regulation today.

How is this tension resolved? Unfortunately, in much of the interaction between NRC and utilities, especially when the schedule for construction of a plant is at stake or when a plant might be turned off because of regulatory actions, the outcome of the conflict is heavily weighted, but not determined by the very large financial costs involved. A plant running at full capacity today generates electricity valued in the range of $1/2 million per day, give or take 50 percent. The daily carrying charges on the funds borrowed to construct a large plant are of the same order. Thus it is very expensive indeed to delay or turn off a large reactor, and all too often a utility will give in to a regulatory demand mainly because the alternative of contesting it is so very costly in terms of construction delays or cost of replacement power. The opposite side of this coin is that sometimes the NRC has been less than vigorous in implementing safety improvements, perhaps because of cognizance that they are quite expensive in many cases.

This is, as I have said, an unfortunate state of affairs. Because the regulatory staff's power to delay or de-rate or close down reactors has such enormous economic consequences, it wields powerful leverage over the utilities. Often, interactions on technical matters are carried out and resolved only at the lower levels of both NRC and the utility, with both sides reluctant to involve higher management, in part because of possible delays. Whether this circumstance adequately protects the public health and safety then depends upon the ability of these lower echelons to agree effectively on the best approach to both safe and economical operation. Again, it is my view that, because of delays and difficulties in getting favorable rulings from the regulatory staff on innovative concepts, this situation is all too often a disincentive because of the way it discourages innovation.

A vital part of the U.S. safety assurance system is the public hearing process, in which at two stages members of the interested public are afforded formal opportunities to intervene in the licensing process. The two occasions are just prior to consideration of an application for a reactor construction permit, and again just prior to consideration of an operating license. (Intervention is possible at other stages, but it is not as often exercised.) These hearings allow the raising of issues involving either safety or environmental impact, and have in recent years been a major platform for the airing of disputes about siting policy. Unfortunately, the hearing process is not as effective as it was hoped to be by the framers of the concept: although one often hears glowing praise for this process as a crucial part (sometimes, it is termed the crucial part) of the public input to reactor decision making, this view is, in my opinion, a vast exaggeration of the impact of the hearing process. This is mainly because by the time a hearing is held most of the difficult issues that separate NRC staff from the applicant have been fought over and resolved by interaction through other channels. Thus the hearing process often finds the staff taking sides with the applicant, against intervenors, perhaps because the staff may bear a psychological commitment to defend the decisions already made that have brought them all to this stage of the process. Also, issues of genuine and difficult technical content seem rarely to be raised in these hearings, although there have been some key exceptions.

No lecture about how reactor regulation works in the U.S. would be complete without mention of the ACRS (Advisory Committee on Reactor Safeguards), an independent statutory body which reviews all reactor plants during the licensing process, reviews operational problems as they arise, advises on the quality and direction of the NRC research program, and generally maintains independent oversight. The ACRS membership consists of dintinguished engineers and scientists drawn from various segments of the U.S. technical community, supported by a staff and backed by the prestige of a quarter century of major impact on reactor safety. Sadly, the ACRS has been more reactive (to staff actions) than it perhaps should have been, in part because of perceived stronger staff competence. However, the ACRS' recent vigorous effort in understanding the accident at Three Mile Island and in overseeing the large NRC research program are a major turn-back toward the crucial role it played in earlier years.

There also exists a vigorous research community working in and concerned about reactor safety: the annual NRC research budget of roughly $200 million maintains this community in the U.S. along with smaller support from other sources such as the Electric Power Research Institute. An equally active community, of comparable size, is working on water reactor safety in other countries besides the U.S., with excellent international cooperation. From this community is drawn much of the expertise that educates, mans, and invigorates all the rest of the nuclear endeavor.

An important part of maintaining safety is feedback from operational experiences, so as to enable corrections to be made in the light of problems as they arise. In this aspect, the U.S. record has not been a good one. Neither the regulatory agency (NRC) nor the various industry groups has maintained an effective mechanism for studying, analyzing, and disseminating the lessons that might be learned from operating experience; and the lessons are sometimes not learned at all, or only on one or a few reactors instead of industry-wide. This dismal situation is in the process of rapid improvement in the light of the accident at Three Mile Island, which was preceded by precursor events whose significance was missed by almost the whole U.S. safety community. If the recently established efforts are as successful as they deserve to be, this aspect at least will be much improved soon.

Nevertheless, there still remain some important barriers to progress in reactor safety. I have already mentioned that in some quarters the concept that reactors are already "safe enough" has prevailed. Also, because of the expense of reactor downtime, some retrofits are prohibitively expensive, or at least are perceived to be so. Sometimes, the development of industry standards can be painfully slow, because of the concensus system used in their formulation. Finally, the government supports almost no research itself at present aimed mainly at improving the safety of light-water power reactors.

This problem stems from what I believe has been a mistaken response after the 1975 Energy Reorganization Act, which created both the NRC and what is now the Department of Energy (DOE) from what had been the former U.S. Atomic Energy Commission. NRC cannot do developmental work because of its regulatory role, and DOE has put most of its own efforts into fast breeder reactor development.

Lest this long litany of problems leave the impression that our regulatory system is ineffective, it is important to point out again that it has many important strong points too. The most important measure of its effectiveness is surely the remarkable safety record of the commercial power industry that is regulated by the system I have described. Even including the unfortunate and significant accident at Three Mile Island, the overall commercial reactor safety record has been outstanding. There have been no large releases of radioactivity, no core melt accidents, no radioactivity-induced prompt fatalities, and excellent control of routine emissions through the two-decade history of over 400 reactor years of U.S. commercial operation. This record, on its face, says a good deal about the overall efficacy of the program. The fact that the record is not strong enough on its own accord to tell us <u>all</u> we want to know about reactor safety should not obscure the fine record just mentioned. Despite the Three Mile Island accident and the many problems it has revealed (and surely after profiting from the lessons of TMI), we do have a good foundation upon which to build the kind of improved safety assurance system that we all desire.

THE ROLE OF QUANTITATIVE RISK ASSESSMENT IN REACTOR SAFETY

Robert J. Budnitz
U.S. Nuclear Regulatory Commission
Washington, DC   20555

In my first lecture, I discussed the way that safety is implemented and assured in the reactor enterprise in the United
States. Now I will discuss quantitative risk assessment and its
role in reactor safety.

First, what is risk assessment? I refer to a set of methodo-
logies that, used together, allow the quantitative analysis of the
risks from a given undertaking, in our case from the operation of
large commercial nuclear reactor power stations. Usually, one
thinks of the risk in terms of the consequences from large accidents, or alternatively in terms of the consequences combined with
some likelihood of occurring. In reactor safety, it is usual to
define "risk" to be the product of consequences times their probability of happening; but hardly anything that I shall discuss
here depends upon that particular choice of definition. Also,
most people generally think in terms of rather large risks, which
means either quite large probabilities or quite large consequences,
or both. That is, most people are not very interested in accidents
with small risks, nor in small accidents with such high frequencies
of occurrence that they already occur in our own experience quite
often. Of course, it is fortunate that generally the most frequent
accidents have small consequences, a fact that we know both from
the safety record of our large reactor endeavor up to now and from
analyses of how accidents happen.

Risk assessment means the quantitative analysis of risk, and
in the public mind that generally brings to mind the study of overall risks, namely what the total risk is from a particular undertaking. Perhaps the most famous example of such an overall risk

analysis is the Reactor Safety Study, also known as the Rasmussen Report[1] or WASH-1400 after its report number.

A key point that I wish to make is that risk assessment means far more than the study of overall risks, at least to me. Indeed, as I will discuss later, in my opinion its most valuable application is not in overall risk analysis, but in more limited analysis and comparison of systems, accident sequences, and study of uncertainty. It is already, and in the coming years will become even more, a tool for improving the design and operation of our large nuclear reactors.

But another prominent feature of risk assessment today is the great controversy surrounding it. This controversy continues in the face of almost overwhelming evidence as to the importance and usefulness of the technique. Why?

The reason, of course, is the Rasmussen Report. Because WASH-1400 was oversold by some proponents of nuclear power, and strongly criticized by many opponents, it became immediately after its publication in 1975 one of the battlegrounds on which the nuclear debate unfortunately has focused. I will come back a little later in this lecture to the achievements and limitations of WASH-1400 itself, as I see them, but here I want to make the point that the controversy has been a major factor in clouding the issue as to the actual value of quantitative risk assessment methods. To me, this is profoundly unfortunate.

Quantitative risk assessment methods are intrinsically difficult to develop and use: they are highly technical, require skill in both engineering and analysis, and even at best yield results that have significant uncertainties. This too has made the techniques controversial. Also, because much public attention has been focused on the Rasmussen Report's conclusions on the "overall" risk from reactor operations, the controversy has tended to neglect the significant applications of these methods for purposes other than overall risk quantification.

The Rasmussen Report was a major intellectual achievement, a breakthrough that has already had great impact. What was the WASH-1400 study? It was a multi-million dollar effort over 2 years (1973-1975), financed by the U.S. Atomic Energy Commission and completed after the AEC's reorganization by its successor agency, the U.S. Nuclear Regulatory Commission (NRC). It attempted to analyze the risks from two large American reactors, one a PWR (pressurized water reactor) and one a BWR (boiling water reactor). The goal was to quantify the probabilities and consequences of large accidents, accidents with significant public consequences in terms of radioactive releases leading to disease, death, and property damage. The report did not study sabotage scenarios.

The report used the methods of event-tree/fault-tree analysis; indeed, in a real sense it helped to pioneer these methods. The study team assembled engineering and mathematical models of a number of processes involved in reactor accidents, including processes of core melt, radioactive dispersion, transport in the environment, and disease/death. Because it was the first major effort of its kind, it had to do much original research in studying some of the areas outside of the reactor itself, such as in radioactive dispersion, core melt, and related phenomena.

But the key breakthroughs occurred in the event-tree/fault-tree approach. In this talk, I cannot go through how this event-tree/fault-tree methodology works. I assume that you are aware that the method treats accident sequences by first assembling various logical diagrams describing accident progression, and then analyzing what failures, in what components and with what time order, lead to the outcome under study. Event tree generation, while difficult, is relatively easy compared to fault tree generation; but each is an excruciatingly tedious process of detailed study of the reactor configuration and the interrelationship of its many functions and parts. The fact that these techniques succeed as well as they do is remarkable.

Of course, the WASH-1400 study has some limitations: these have been brought out in the intervening years by a large number of critical articles and memoranda, and were studied in 1977-1978 by a special Risk Assessment Review Group, under the chairmanship of Professor H. Lewis, put together by the NRC for that express purpose. The Group's Report[2] is a good reference on many of the issues I will cover in this talk.

In my opinion, the main limitation of the Rasmussen Report was that in retrospect it was too ambitious: the study team attempted to put numerical values on everything they needed to analyze, and in some cases the underlying data base was weak or non-existent. Also, some of the engineering models, in particular the models for seismic-induced accidents and for fires, seem weak. The way human intervention was modelled was inadequate, a fact that the study team itself recognized: that is, although human error was put into the accident sequences wherever possible, there was an inadequate treatment of the possibility that the operators might cope effectively during accidents, thereby mitigating their otherwise large consequences. Also, the way that human error might compound an accident was inadequately treated, an observation that is highlighted in the light of the accident at Three Mile Island.

Other weaknesses in WASH-1400 were in the radioactivity dispersion model, the latent cancer model, and in some of the data
base. Finally, the report was hard to read and had an atrociously
poor Executive Summary. Still, despite these identified weaknesses,
WASH-1400 stands as a monument in the development of quantitative
risk assessment methods.

Figures 1 and 2 show the overall results of WASH-1400 as published in its Executive Summary. The figures show on the abcissa
a measure of consequences, in this case prompt fatalities from
accidents studied in the analysis. On the ordinate are shown the
probability results, and the curves reveal, on their face, that the
risks as measured by prompt fatalities from the two reactors studied
are quite low compared to a whole range of other risks that are confronted by our advanced society. (There are other consequences from
reactor accidents besides prompt fatalities, which I am not discussing here. The most important are latent cancers and property damage.) The fact that the uncertainties in the WASH-1400 analysis
are now perceived to be quite large should not obscure the general
conclusion that the risks of these two reactors are likely to be
reasonably small.

I will not discuss these figures further: their study is of
some interest to anyone who has not seen them before, but has little
interest now to the reactor expert.

I wish instead, in the remainder of this talk, to concentrate
on what to me is a more important issue: that is, the more valuable uses of these techniques for analyses other than overall
risk assessment. I will use examples from applications within my
own agency, the NRC.

In the NRC, our regulatory staff is faced with almost count-
less occasions in which decisions have to be made about various
issues. Examples include decisions about whether a flaw in a
particular safety system is serious enough to merit prompt remedial action; whether the resolution of a particular "generic"
safety issue is of high or low priority; whether a modification
in an otherwise standard design is, on balance "safer" than its
predecessor; whether a certain new regulatory requirement should
be required for retro-fit on all existing reactors, or required
only on reactors still in the design stage; and so on. These
decisions are the day-to-day heart of our regulatory process, and
the establishing of priorities among the many things the NRC staff
could do with much larger resources is vital, since our resources
are limited.

This is where quantitative risk assessment methods come into
play. They allow the analyst to cast light on how significant the
risk reduction potential is from certain actions; how important

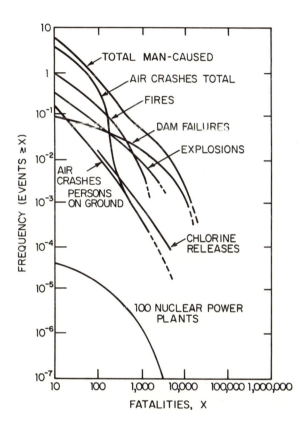

Figure 1. Frequency of Man-Caused
Events Involving Fatalities.

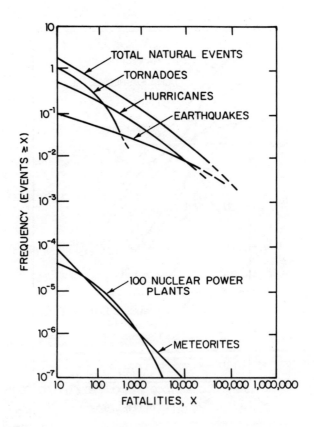

Figure 2. Frequency of Natural Events
Involving Fatalities

the risk implication is from various new and unstudied accident
sequences; and so on. In short, these techniques are a way to
rationalize the process of studying and regulating, to concentrate
resources where they will likely have the most impact in safety
improvement or safety regulation.

Some recent example of applying these techniques will demonstrate the point. First I will discuss the issue of "generic safety
issues," which are defined in our regulatory process as issues that
apply across a broad spectrum of reactors, rather than only to one
specific design or one reactor, and are unresolved in the sense
that no firm regulator position has been taken on them. Last year
our agency informed our Congress of 133 of these "unresolved generic issues." This year, the list of important ones was reduced to
only 19! This reduction was not accomplished by the "resolution"
of all of the others, but by an analysis showing that only the 19
remaining had enough real or potential safety significance to merit
concentrated study over the near-term. The others either had small
risk significance, or in some cases had no actual risk significance
at at all. This has served to allow our staff to concentrate its
effort on the _important_ items, leaving to some future time the study
of the others.

Another recent application of these techniques occurred just
after the accident at Three Mile Island. I will remind the audi-
ence that one part of the TMI accident involved the unintentional
blockage of some valves that prevented auxiliary feedwater from
supplying water to the large steam generators after the regular
main feedwater pumps had tripped. Although we now believe that
the unavailability of auxiliary feedwater (AFW) for the first few
minutes of this particular accident did not materially affect the
accident outcome, our agency became concerned that perhaps there
might be a general problem with the availability of AFW systems in
other pressurized water reactors.

To study this question, our NRC staff, in collaboration with
some colleagues elsewhere in the country and with the reactor vendors and utility owners, undertook a quick study of AFW reliability,
just after the TMI accident. Thirty-three PWRs, representing 25
different AFW configurations, were studied. The AFW systems typically involve a set of redundant and diverse pumps (some electric
driven, some steam driven). The reliability of the whole AFW system then depends upon their geometrical configuration, the location
of various valves that might be mistakenly closed, the source of
water supply for these pumps, their control systems, and so on.
The study attempted to do a relative comparison of availability
of AFW systems under normal conditions, under conditions of loss
of offsite power (LOOP), and under conditions of loss of all AC
power, which would include loss of the onsite power.

The tentative results of the recent reliability study are shown in Figure 3. The more reliable AFW systems are toward the bottom of the figure, and the horizontal scale is only approximate, with about one order of magnitude of reliability shown by the arrows near the left bottom. The figure has a remarkable message: namely, that in ordinary operation the reliability of these AFW systems varies from one PWR to another by almost two orders of magnitude, and that some reactors have very different reliabilities after loss of all AC power than after loss of only offsite power!

Without dwelling on the details of this figure, I wish instead to dwell on its message: the use of quantitative risk assessment methods is a powerful tool indeed for comparisons such as these! The insights enable the regulatory staff, and the utilities, to concentrate effort where it is needed most.

Other examples of application of these methods are in helping to develop emergency plans; in developing a rational basis for better regulation of transient and small-LOCA events; in studying possible weaknesses in our older reactor plants; and in allocating resources to study operational data. All of these applications are now underway or soon will be.

Let me summarize my views on the future of risk assessment in reactor safety as follows: I believe that these methods already are beginning to play an important role in making our regulatory process more rational within NRC. I believe that their importance will be even greater with each passing year over the next few years, in part because the still small number of practitioners of the art of quantitative analysis will be increasing, but mainly because so many people now believe in their use. I believe that the techniques will be most useful in making comparisons of similar systems (such as auxiliary feedwater systems), or next greatest help in studying safety significance of particular accident sequences, and less useful for overall risk assessments.

I will elaborate on this last point briefly: to make an overall risk assessment for a particular large facility requires that the analyst do a complete and accurate analysis that covers all of the significant contributors to risk. This will always be difficult even in the best of circumstances, and will have large uncertainties. Nevertheless, I believe that the techniques have a role to play in overall risk assessment as well. After all, even with their flaws, these quantitative risk assessment methods are the only quantitative methods we have--and I hope and expect that we will make the most of them.

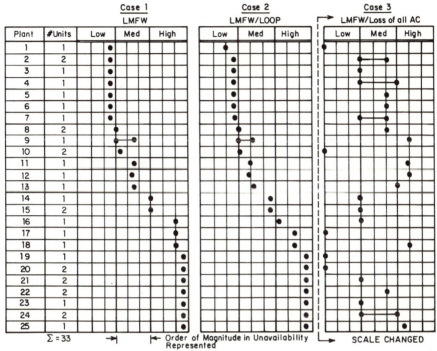

Figure 3. Comparison of Reliability of 25 Configurations
(in 33 PWR Plants) of Auxiliary Feedwater Systems.

REFERENCES

1. The Reactor Safety Study, Report No. WASH-1400, U.S. Nuclear
     Regulatory Commission, Washington, DC (1975).
2. H.W. Lewis, et al., NUREG/CR-0400, "Risk Assessment Review
     Group Report to the NRC," Washington, DC (1978).

COAL CONVERSION : ADVANCED TECHNOLOGIES

Burkhard Bock

University of Essen
43 Essen 1, Germany

SUMMARY

Coal has been refined since the substituting fuel-wood
to heat by combustion or to coke by devolatilization. Using the
heat produced by combustion coal is converted to electrical
energy. Finally via coke iron is produced by the reduction of
iron ore.

In the last decades coal has been pushed aside by petroleum and natural gas in the market of energy and raw material,
because these are easy to handle. With the increasing scarcity
of the natural fluid and gaseous energy carriers coal has to be
substituted for the former competing substances - and possibly
in the same state of aggregates.

Converting to heat, coke, oil and gas are both new and old
tasks of coal-technologies. For many years, especially since
the "oil-crisis", new techniques have been tested thereby the
coal should be refined more economically and enviromentally
friendly. Such technologies are among others as follows:

- Carbonization with coal-preheating or by for-
med-coke-processing

  Gasification   in fixed bed, fluidized bed and
  entrained bed with H_2 or H_2O by
  using nuclear reactor's wasted heat

- Liquefaction   by hydration or gas synthesis after
  coalgasification

- Combustion   in the fluidized bed combined with
  steam- and gas-turbines.

COAL AND ITS PRODUCTS

Coal as raw material and source of energy

Coal, like petroleum and natural gas, is not only an energy source for heat production and power generation just as the tasks of nuclear fuel, sunradiation or the waterpower are. Coal is the most carbonacious raw material for chemical processes. It is followed by petroleum and natural gas with decreasing contents of carbon and increasing contents of hydrogen. (Fig. 1)

	C	H	O	N
HARD COAL	75 - 97	2 - 5	2 - 2o	o - 1 w-%
PETROLEUM	8o - 88	1o - 14	1 - 7	o - 1
NATURAL GAS	5o - 71	14 - 24	1 - 2	1 - 7

Fig. 1.   Elementary analysis of coal and its competition-fuels

Oil and gas are raw materials, easily transportable in pipe-lines, less difficult to dose in refiningplants, by flowing through reactors and in reactions with other solid materials, liquids and gases. Therefore it is not surprising, that petroleum and natural gas have dominated the energy market in the last decades.

In West-Germany the portion of these liquid and gaseous materials in the energy market rose from 5 % in 1950 up to 68 % in 1978 (Fig. 2). Coal was diplaced of its position from 88 % to 27 % in the same period of time.

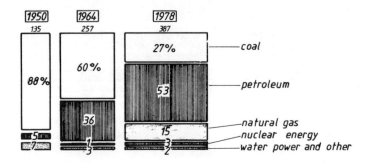

Fig. 2.   Development in Energy Market of the FR-Germany 1979

Should and must we return from oil back to coal and the
other energy carriers in the future the industry wouldn't be
the only one, exchanging raw materials for its large plants.
In 1979 40 % of the oil consumption in West Germany was in
private hands (Fig. 3). This large amount was used for heating purposes (22 %) and for private automobiles (18 %).

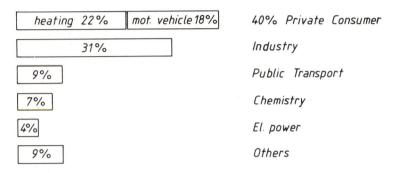

heating 22%	mot. vehicle 18%	40% Private Consumer
31%		Industry
9%		Public Transport
7%		Chemistry
4%		El. power
9%		Others

Fig. 3.   Consumers of Petroleum in the FR-Germany 1979

Everybody knows lately   since the second petroleum chock
in this year   that petroleum is not only expensive but also scarce.
The last world energy conference brought new data about availability and consumption. If we allow the energy stock in the form
of mineral resources to be 100 %, than we have 41 % in form of
coal and 54 % uranium, but only 5 % petroleum and naturalgas
(Fig. 4).

If we extrapolate the consumptions up to the year 2000 (not
taking into consideration how political influences and changing
thought of consumershabits affects the curves course), only
petroleum and naturalgas consumptions would make up 70 % of
all energy-carriers from the soil (45 + 25). There is an extreme disproportion between stockage and consumption. As a result of this, 61 % of the source of petroleum and 54 % naturalgas
would be consumed up to the years 2000 .

Contrary to this, hardcoal and uranium will lose only 1, 7
and 0, 6 % of their stockages up to this time. Therefore reason
favours the greater usage of coal and nuclear energy. Coal will
be used more and more in that degree as it is transformable
to liquid and gaseous substances, easily handled like petroleum

and naturalgas. Besides this, enviromental friendliness will
play an important part.

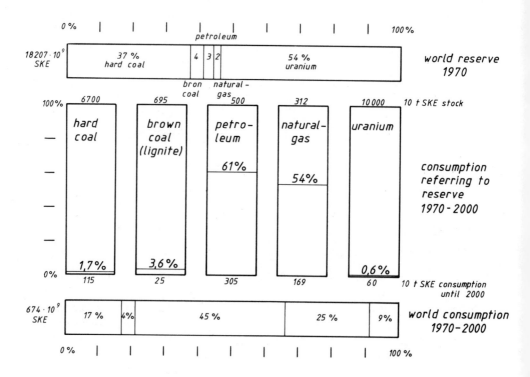

Fig. 4.   World reserves and consumption   (1)

The world wide spread of coal stockages (Fig. 5) serves
coal supply without great problems of transportation all over
the world. Applying suitable refining processes fitted for each
coal type we can use coalproducts like gas, oil, coke, heat
and electricity everywere globally. Allthough this picture looks
so favourable not all coaltypes are of equal quality. Coals of
various behaviours are required in the refining processes . And
coals differ in type, because the original composition is different according to climate and in the world which for million
of years have been different.

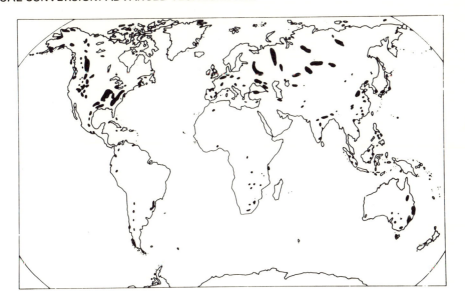

Fig. 5.   Coal deposits of the world (2)

Quality characteristics of coal

Coal is a mixture of organic and mineral substances. It originates mainly from dead plants. This material is enriched in carbon content and becomes lacking in hydrogen and oxygen in the absence of air-oxygen, with increasing time, at high temperature and under the pressure of the overlying layers.

The so called period of "coalification" was the first conversion, a "wood-conversion". It yields a series of young to old coals, for instance peat, brown coal and hard coals with high and low content of volatile matter. (Fig. 6)

COALIFICATION SERIES	SPEC. WEIGHT T/M	HEATING VALUE KJ/KG	WATER CONTENT %	VOL. MATTER % WF	H %WF	O %WF	C %WF
WOOD	0,2-1,3	14 x1ooo	(DRY)	8o	6	44	5o
PEAT	1,0	14 - 2o x1ooo	6o-9o	65	6	34	55-65
SOFT-BROWN COAL	1,2	2o - 24 x1ooo	3o-6o	5o-6o	5	28	65-7o
HARD-BROWN COAL	1,25	24 - 29 x1ooo	1o-3o	45-5o	5	18	7o-8o
FLAMM-FETT COAL	1,3	29 - 33 x1ooo	3-1o	17-45	4	17	8o-9o
ESS-MAGERKOHLE	1,35	33 - 35 x1ooo	3-1o	7-17	3	5	9o-93
ANTHRACIT	1,4-1,6	35 - 37 x1ooo	1- 2	4- 7	2	2	93-98

Fig. 6.   Coalification series

Hard coals have an important property for thermal refining processes: beyond approx.350 ^{o}C they liberate gases
and tars, most of them being softened and forming cake or
agglomerated cake. Beyond 550 ^{o}C and after a certain degasification time the coalmass is solid (coke) hardening
up to 1000 ^{o}C and liberating gases. Also around 1000 ^{o}C
there is an important temperature for converting processes, especilly combustion because of its ashcontent and the
composition of ash, that melts at this temperature.

Fig. 7 shows the softening behaviour of 3 different
coals. In the case of "Gaskohle" we can see a first period of flowing   (Contraction under 420 ^{o}C) and a second of blowing up, originated from the devolatilazition
(Dilatation beyond 420 ^{o}C).

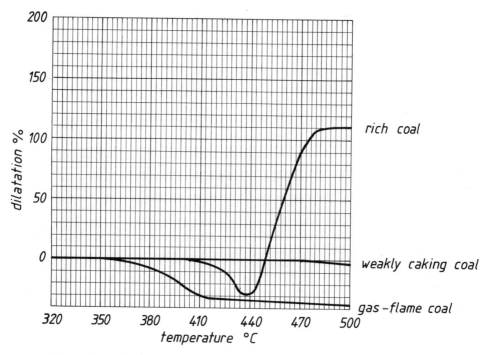

Fig. 7.   Dilatation and contraction of hard coal over
temperature (3)

Refiningprocesses and products (Fig. 8)

process	product	competition
screening + moulding	coal fine grained coal briquette pellet	-
combustion	electricity heat	electricity from nuclear e. heat from nuclear energy
carbonization	lumpy coke fine grained coal formed coke gas hydrocarbons	converted natural gas in iron-ore processes
liquefaction	oil gas extract	petroleum
gasification	syntheses gas methan hydrogen	natural gas

Fig. 8.   Coal refining processes and products

Coal has always been refined. Today refinement starts with purification of crude roughcoal, too: seperation of accompanied stones, division into poor ashed coal, rich ashed coal and stones. For different applications we need different sizes of coal. Coal is grinded or coarsened. For the second case a briquetting or pelletizing process is applicd. A new method is hot briquetting and hot pelletizing, making use of the coals softerning behaviour. The most significant refining process today is coal transformation to electricity as an energy carrier or coke as a reducing agent in ironproduction. Competing with coal, but in cases also supplementary processes, are nuclear-power stations with their electricity- and heat - products. For iron production there is a competing process, whereby ore is reduced by the converted naturalgas.

A further, yet here undescribed field of coalrefining is special coke manufacturing: porous finecoke as a filter aid, coke granulates with a large internal surface as an absorbing agent (activated carbon) and granulated coke with a defined porous radius for gas purification (molecular sieve).

For the future however, the most important are, as described, liquefaction and gasification of coal. Today petroleum and naturalgas are viewed as competitors of coal. For the future they have given scales being easily handy, unlike coal. Fluid products of coal vary from heavy to light hydrocarbons also produced from petroleum by cracking and distillation.

Solid, liquid and gaseous products have always been supplied from the coke oven process, where easily 50 % of hard coal has been refined hitherto. We manufacture out of 1000 kg coal about 775 kg coke, 180 kg gas (365 m^3), 30 kg tar, 10 kg benzene and 2 kg ammonia.

In the future quantities and qualities of coal products will be dictated by the market according to the customs with the fuel today. And so, liquification and gasification of coal have to enlarge the number of coal refining methods again. Again, for there are already old processes, which we have only to improve.

Advanced processes will be objects of discussion in the following lectures:

- Gasefication with steam and hydrogen,
- Liquefaction by hydrogenation or gas-synthesis,
- Carbonization to lumpy, fine grained or formed coke,
- Combustion in the fluidized bed by normal and elevates pressures.

Advanced technologies

Thermal coal refining is almost a process with the praticipators, solid and gaseous materials. The solid material is coal or coke, the gaseous either degasification product of coal or a gaseous reaction's partner. Such a process consists of carbonization, gasification and combustion.

Chemical engineering knowledge offers three possibilities

for such processes, as illustrated in <u>Fig. 9</u>: fixed bed, fluidized bed, entrained bed.

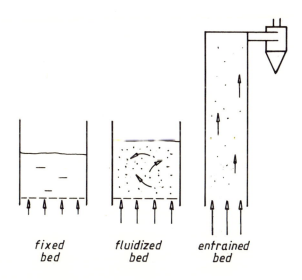

Fig. 9. Technologies for contact of solid
and gas

There are two methods of coal liquefaction:

1. Hydrogenation of coal with molecular hydrogen or a hydrogen-enriched solvent

2. Gassyntheses with catalysts after gas-production by gasification

<u>Figure 10</u> shows the way of coal conversion to gasoline using both above mentioned methods.

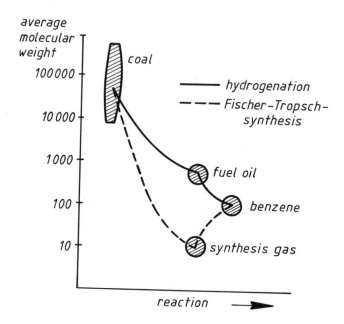

Fig. 10.   From coal to benzine via hydrogenation
            and synthesis (4)

CARBONIZATION

Coal, coke and gases

Carbonization is coal degasefication, that is the delivery of the so-called "volatile matter" (v. m.). Carbonization products are the outgased coal (char) and a gas mixture of high and low boiling hydrocarbon compounds and some other components like oxygen, nitgrogen and sulfur.

The motive force of carbonization is heat, being applied until reaching a high temperature of about 1000 °C. Coal is thereby subjected to series of chemical and physical reactions termed "pyrolysis" all together. Coal with high contents of oxygen like lignite form herewith many oxygen-compounds as carbonmonoxide and carbondioxide. (Fig. 11)

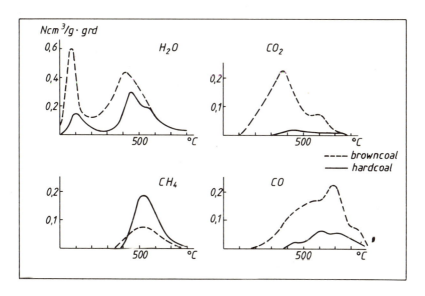

Fig. 11. Formation of different gases from degasing
browncoal and hardcoal (17)

Within the hard-coal-series the tar output increases
anderstandably according to the amount of volatile matter
and at the same time the coke-output decreases. (Fig. 12)

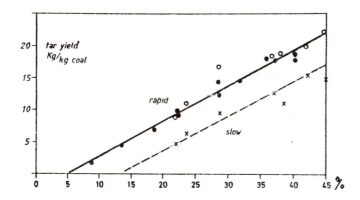

Fig. 12 a. Tar-yield dep. on content of vol.
mater in rapid and normal carbonization (5)

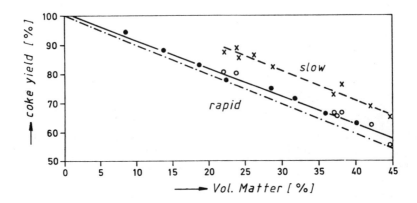

Fig. 12. Coke-yield dep. on
          content of vol. matter in rapid and
          normal carbonization (5)

The heating up velocity has a great influence on pyrolysis. In the coking chamber furnace a heatingup-rate of
around 3 °C per minute can be achieved, but in a quick-
carbonization a rate of up to 10000 °C per minute. The latter
yields a large quantity of tar. This, when coal is slowly heated, liberates many gases.

Volatile matter and heat-up-velocity strongly influence
the cokes quality. (Fig. 13 and 14)

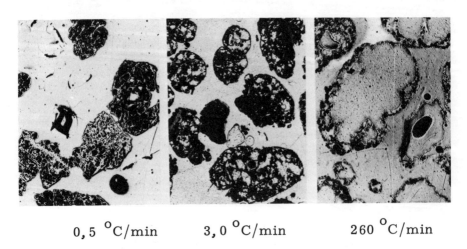

0,5 °C/min   3,0 °C/min   260 °C/min

Fig. 13. Pore formation of fine grained coke dep.
           of heating rate (Vol. matter 22,3 %) (6)

Vol. M. 8,4 %

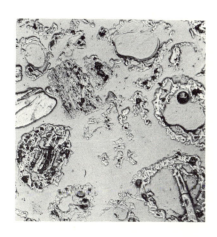

Vol. M. 13,8 %

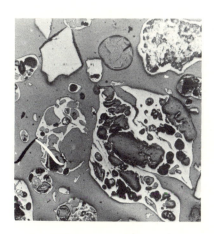

Vol. M. 21,2 %

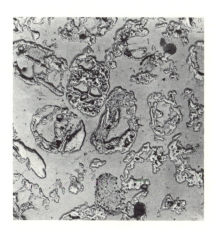

Vol. M. 38, 2 %

Fig. 14.   Pore formation of fluidized bed coke, dep.
on the content of vol. matter of the coal (7)

Heating quickly means normally heating single grains by gas or a solid heat carrier. By the possibility of free expansion porous cokegrains originate directly from coalgrains under the inner pressure from the quick escaped gases. Heating slowly is practised in an indirect heated coke chamber or in a flash gas oven. The coke oven process uses the good softening and caking conditions of the so called "medium volatile coking coal" in shaping irregular broken pieces. In the flash gas oven only non coking coals are carbonized, heated by a circulating hot gas.

Principles of carbonization

The most realized   principle to date is the indirect heating of a coal bed within the closed walls of a coke chamber. (Fig. 15 above)

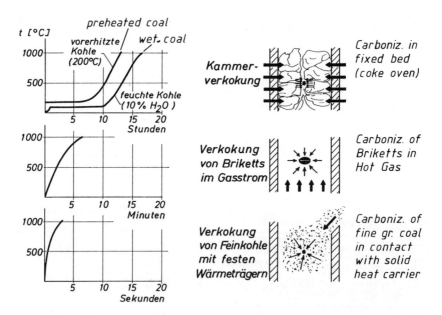

Fig. 15. Trend of Temperature in different
carbonization processes

An opposite process to this is the quick carbonization and conversion to fine grained coke by gaseous heat carrier in an entrained bed, by a solid heat carrier in a solid-material-rotating-process or by a solid and gaseous heat carrier in a fluidized bed-prozess. (Fig. 16)

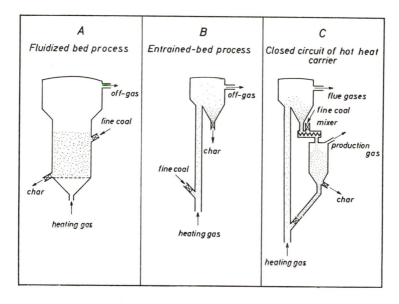

Fig. 16. Processes for production of fine
grained coke (rapid carb.)

The differences between heating up and carbonzation among these processes are connected to differences in products: lumpy coke and fine grained coke. From the latter moulded cokes are formed. And there are differences in coal bases: caking coals for lumpy coke, non caking coal for fine grained coke or formed coke.

Production of lumpy coke

The chamber oven process, well - known since the beginning of the 19th century, has been improved during the last few years. A higher coal charging rate was achieved by using more solid stone materials of increased thermoconductivity in building thinner chamberwalls and by performing enlarged furnace rooms. In all these the preheating of coal charge played a very significant part. (Fig. 17)

Regarding enviromental pollution exhausters for the filling process and central cooling plants for hot coke were installed.

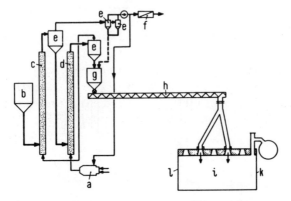

Fig. 17. Flow sheat of "Precarbon-
Process" for preheating
coal to coke even (8)

Production of fine grained coke

    Examples of fine-coke-production according to the three
above mentioned principles are:
- the fluidized bed carbonization and the LR-solid heat-
  carrier process as processing steps used in formed-
  coke-experimental plants of Bergbau-Forschung
- the entrained bed process of Bergbau-Forschung (Fig. 18)

    In all cases fine grained coke is obtained direct from
fine grained coal. However a caking of the grains to agglo-
merates is unwanted and must be prohibited.

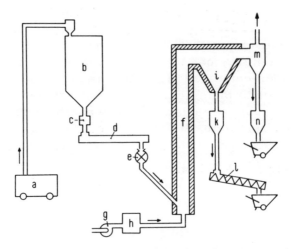

Fig. 18. Devolatilization of fine grained
coal in an entrained bed process (8)

Production of formed coke

Reasons for the production of formed coke are to be found in the intention of having a widespread coal basis, a improved heat-transfer and the production of an improved blast-furnace-coke. The processing stages are usually predegasification of fine coal, its mixing with a binder, the moulding and the part degasification of the formed pieces afterwards. (Fig. 19)

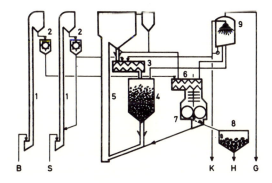

Fig. 19.   Scheme of BFL-plant "Prosper"
for the production of formed
coke with LR-charmaking (5)

In many experimental plants predegasification is practiced by entrained bed heating or in contact with a solid heat carrier or in fluidized bed carbonization. The forming process is a briquetting or pelletizing step. And the part degasification afterwards takes place in flashgas or in contact with a solid heat carrier. Fig. 19 + 20 show flowsheets of 2 plants of Bergbau-Forschung for the production of hot-briquettes by different manners of making char.

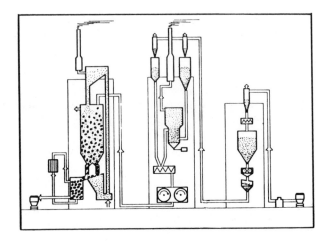

Fig. 20. Scheme of BF-plant for the production of formed coke with
fluidized bed char making (7)

GASIFICATION OF COAL

Gas from coal

Since the beginning of the industrial age coal has been
source of energy production and a feedstock for chemicals.
Starting in the 30's first oil and then natural gas came into
the market and are still dominant today, althoug in some
countries coal is still the basis for energy. (Fig. 21)

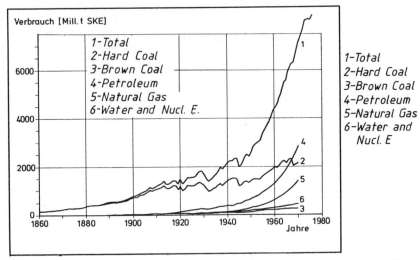

Fig. 21. World consumption of energy carriers (9)

In the field of secondary energies gas has a lot of advantages: low emissions to the enviroment during utilization, economic transmission and distribution, possibility of storage and many-fold uses.

The extent to which the private energy consumer values easy to handle fuel, becomes evident in data from West-Germany, which are shown in Fig. 22.

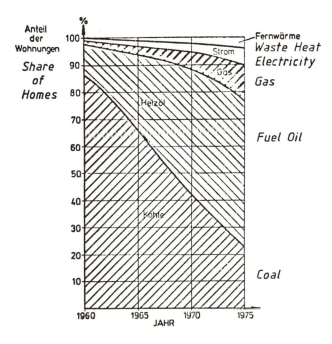

Fig. 22. Energy for room heating (9)

According to that figure coal consumption for heating rooms diminished from 86 % to 22 % between 1960 and 1975 meaning a decrease to a fourth, while the consumption of fuel oil and gas increased from 11 to 54 % and from 2 to 12 %, a five to six fold increase.

During the energy crisis in 1973 the private consumer did not change the fuel basis on account of the technical difficulties, while the industrial consumer was able to use more coal instead of oil.

Fig. 23 shows, that at this time oil consumption of the industry was significantly reduced in contrast to coal consumption.

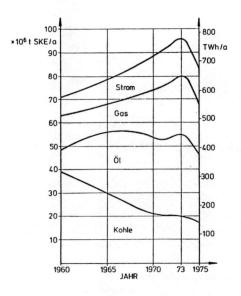

Fig. 23.   Industrial consumer
during energy crisis (9)

Fig. 24 gives an impression of transport costs, comparing that of fine and briquetted brown coal with current (usually won from brown coal) and methene, the main component of natural gas. We can see, that the transport costs of gas, related to enthalpy, lie far below that of current.

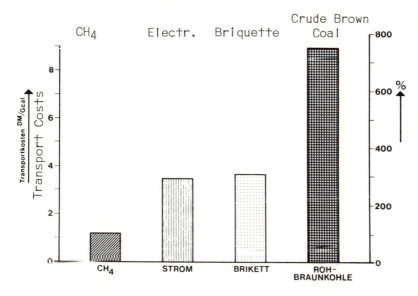

Fig. 24. Cost for transportation of 1 Gcal over 100 km (1)

Regarding these aspects the manufacture of gas from coal by gasification if possible in combination with nuclear reactors seems to be very attractive and can play an important role in future energy technology worldwide .

Gascomposition and -utilization

If we want to develop new technologies of attaining gas from coal, it is necessary to consider some gases on the market and to compare them with each other. (Fig. 25) Industrial and private consumers today have at their disposal mostly the low caloric gas from blast furnaces, the degasification product of chamber oven and natural gas.

While blast furnace gas,because of its low calorific value is only used by direct consumption in power stations or for air preheating, the other two gases conveyed in pipelines and used by consumers for many heat-and power processes. Gases from coke oven and natural gas are suitable for chemical refining processes, because they contain appreciable concentrations of hydrogen and methene. Special gases for this purpose are synthesis and reduction gas, attained by special gasification processes. They contain large quantities of carbon-monoxide, hydrogen or methene.

	Schwachgas	Wassergas	Synthese- oder Reduktions- gas	Luigi- Ruhrgas	Koppers- Totzek	Stadtgas	Erdgas	Gasart	
CO	29	40	12	55	5,5	—			
H_2	10,5	50	43	30	54,5	—			
CH_4	—	—	11,5	0,1	25,3	87,5		Zusammen- setzung	analysis
CO_2	5,5	5	32	13,2	2,3	3,1			
N_2	55	5	0,3	1,4	9,6	0,6			
	%	%	%	%	%	%			
	4600...12500 KJ/Norm m^3	ca. 12500			16700... ...20000	25000... ...37000		Heizwert	caloric value
	Low Calorie Gas	Gas for synthesis		Gas for reduction	Gas from coke oven	Natural Gas			

Gases for heating, Gase für Heizung, power station, Kraftwerk und Synthese and synthesis

Fig. 25. Analysis and caloric value of gases for
heating, power staion and synthesis

The fundamental reactions of gasification

Gasification is the reacton between the solid organic materials and a gasifying agent, in which coal is completely converted to gas leaving the ash   . Althoug coal has a very complex molecular structure it is possible to discuss the basic reactions regarding only carbon. Fig. 26 shows four primary reactions between the solid and different gasifying agents and two secondary reactions which occur between the gases formed and the gasifying agent in the gaseous phase. In existing gasification processes steam is of utmost importance. Its reaction with the carbon leads to CO and H_2 and is strongly endothermic.

		ΔH	
Primary reactions:		$\dfrac{Kcal}{mol}$	% of heat of combustion
Water gas reaction	$C_{solid} + H_2O\,steam \longrightarrow CO + H_2$	+28,3	29 endo.
Boudouard reaction	$C_{solid} + CO_2 \longrightarrow 2CO$	+38,3	40 endo.
Partial combustion	$C_{solid} + 1/2\,O_2 \longrightarrow CO$	-29,4	30 exo.
Hydrogasification	$C_{solid} + 2H_2 \longrightarrow CH_4$	-20,9	22 exo.
Secondary reactions:			
Shift reaction	$CO + H_2O\,steam \longrightarrow H_2 + CO_2$	-10,1	14 exo.
Methanation	$3H_2 + CO \longrightarrow CH_4 + H_2O\,steam$	-49,2	51 exo.

Fig. 26. Basic reactions of coal gasification

Another gasifying agent is CO_2, which interacts with the solid carbon in the so-called Boudouard reaction to give CO. It is even more endothermic than the water gas reaction and plays an important role for instance in the blast furnace, but is of minor meaning for technical gasification. The reaction of carbon with oxygen, the combustion, also can be summarized under gasification. Specially the partial combustion $C + 1/2\,O_2 \longrightarrow CO$ which is exothermic is used in combination with the water gas reaction to fill the heat requirement. Finally the reaction of C with H_2 leads to CH_4, this reaction is exothermic and is a fundamental process in new developments of coal gasification with respect to SNG production. As secondary processes, in the gaseous phase there are the shift reaction in which CO and H_2O react to H_2 and CO_2 and the methanation reaction in which H_2 and CO are converted into CH_4 and steam to be rentioned. Both are exothermic and play an important role in technical gasification reactors and in processing steps of the raw gas as shown afterwards.

The composition of the produced gas depends upon the processing steps and conditions wich are chosen corresponding to its end use. Gasification with steam is strongly endothermic and that with hydrogen exothermic. In the former case mostly hydrogen and carbonmonoxide are produced, suitable for the reduction of iron ore and synthesis in the chemical industry. The latter leads almost to methan the basic component of natural gas, used specially for heating purposes.

Fig. 27 gives values of gas analysis at thermodynamic equilibrium depending on temperature at a pressure of 1 bar and pressure at a temperature of 800 $^{\circ}$C as calculated by thermodynamic rules for steam gasification.

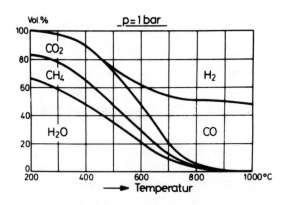

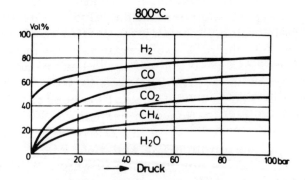

Fig. 27. Equilibrium of gas from steam gasification (11)

Principles of gasification technology

For energy production in future steam gasification and the gasification with hydrogen - hydrogasification - will be important, whereby the first one is a basic process as schematically shown in figure 28.

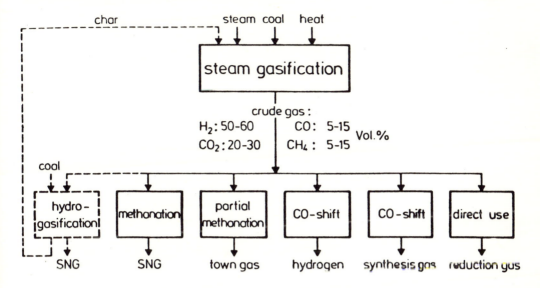

Fig. 28. Processes and products of coal gasification (12)

Steam and coal react in a strong endothermic process at temperatures above 750 ^{o}C, whereby the organic part of the coal is transformed to a crude gas which is rich in hydrogen but contains also CO, CO_2, and CH_4. The crude gas after cleaning can be used either directly for instance as reducing gas or by an adjacent combustion for clean energy production. It also can be transformed by well known processes to synthesis gas, hydrogen, town gas or to CH_4, as a substitute natural gas (SNG). For the latter purpose two ways are possible: methanation or more advantageously hydrogasification as will be shown later on.

Steam gasification of coal is a well known technique which has been performed world wide on an industrial scale in many variants. As shown in figure 29, in conventional processes the heat requirement of the endothermic reaction is supplied by burning part of the coal in the gasifier with oxygen. Additional coal has to be burned in a power plant for the production of the steam and electricity needed in the process. The concept of nuclear steam gasification has the aim to use a high temperature nuclear reactor (HTR) as a heat source to supply the heat needed in the gasifier, for steam production and for other requirements of the plant.

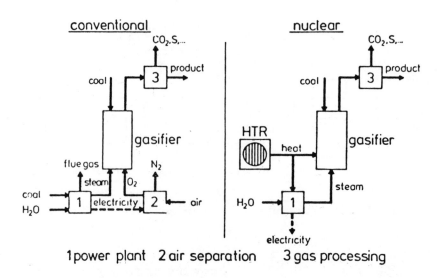

Fig. 29.   Conventional and nuclear steam gasification
of coal (12)

Thus coal only serves a raw material for the gas to which
it is transformed optimally and no oxygen is needed.

Figure 30 shows simplified flow sheets for hydrogasification which is an exothermic reaction. However, heat is
needed for the production of hydrogen. There are two possibilities for its production. The first one corresponds to the
conventional processes which are under development especially in the USA.

As shown on the left a hydrogasification is combined with
a steam gasifier so that the residual char from the hydrogasification is used for the production of hydrogen in the steam
gasifier. This combination is attractive as the reactivity of
a char against hydrogen decreases strongly with the bum off
so that the total char cannot be gasified in the hydrogasifier.

The reactivity of this char against steam, however, is very much higher than that against hydrogen resulting in an economical use of the char in the steam gasification. Nuclear heat can be used in this process replacing conventional steam gasification as shown in the previous picture.

The second possibility to use neclear heat is shown on the right: The CH_4 formed in the hydrogasification step is partly used as SNG and partly fed to a methane reformer in which it is converted into H_2 and CO by a catalytic steam cracking, whereby the heat necessary for the reaction and steam production is supplied by the HTR. The hydrogen, possibly after a cleaning, is fed to the hydrogasifier where it react with the coal. The residual char has to be used for other purposes.

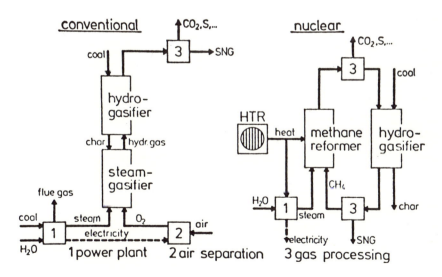

Fig. 30. Conventional and nuclear hydrogasification of coal (12)

Developments
- - - - - - - - - - - -

Frequently the chemical engineer is confronted with installing processes whereby a solid material reacts with gas
while heat must be added. This is the case in the gasification
process with steam.

Figure 31 shows five of these variants. The first has long
been known as autothermal process: heat is produced by partial
combustion of coal with oxygen within the layer. There coal is
not fully gasified, but also undergoes a combustion. Therefore
the process gas a mixture of gasification and combustion gas.

The other processes do not derive heat from coal, but
from other sources. In the cases 2 and 5 heat is indirectly
transfered from reactor walls or immersed tubes to the layer.
The last process (5) illustrated corresponds to a concept of
waste heat transfer from nuclear reactor to a gasifying fluidized bed. The third case is the so called "circulation of solid
heatcarriers": coke is heated during back feeding from the
plant's base to the top and delivers the heat necessary for
gasification while adding to the feed coal. In the fourth case it
is the gas which is delivered back and heated again.

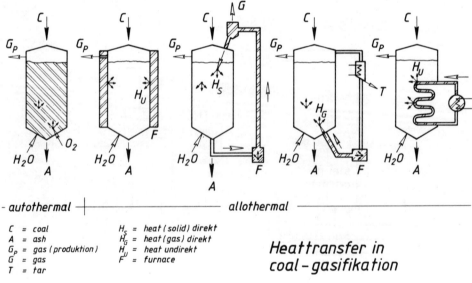

C = coal H_S = heat (solid) direkt
A = ash H_G = heat (gas) direkt
G_P = gas (produktion) H_U = heat undirekt
G = gas F = furnace
T = tar

Heattransfer in
coal-gasifikation

Fig. 31. Heattransfer in coal gasification

Variants of autothermal gasification are illustrated in Figure 32. There are three possibilities of reaction between solid material and gas, as they are handled in combustion and many other processes:
fixed bed, fluidized bed and entrained bed. The additionelly temperature and velocity curves make it obvious, that fluidized bed is operating at middle and relative constant values. Very high temperatures exist in entrained bed processes, low re- locities in the fixed bed.

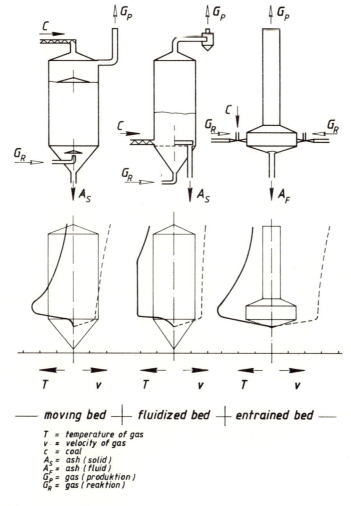

T = temperature of gas
v = velocity of gas
c = coal
A_S = ash (solid)
A_F = ash (fluid)
G_P = gas (produktion)
G_R = gas (reaktion)

Fig. 32.   Process for steam gasification
          of coal

Examples of technical plants according the mentioned principles are shown in Figures 33 to 36.

Fixed bed with gasification- and combustion-gases flowing through the filling; outlet of gasificatin gas on the left; that of devolatization on the right. (Fig. 33)

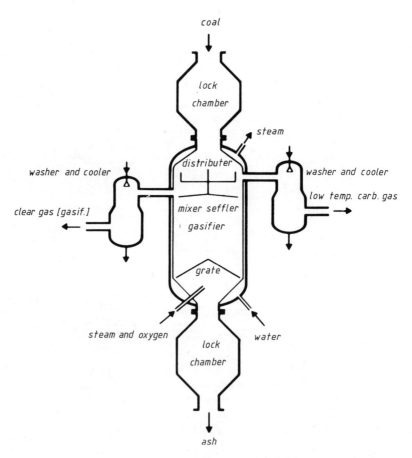

Fig. 33. Lurgi - pressure - gasification (13)
(Typ Ruhr 100)

Fluidized bed plant with up-whirling of the layer by gasification and combustion gases; feeding gases and coal from the right; recycling of discharged coke on the left. (Fig. 34)

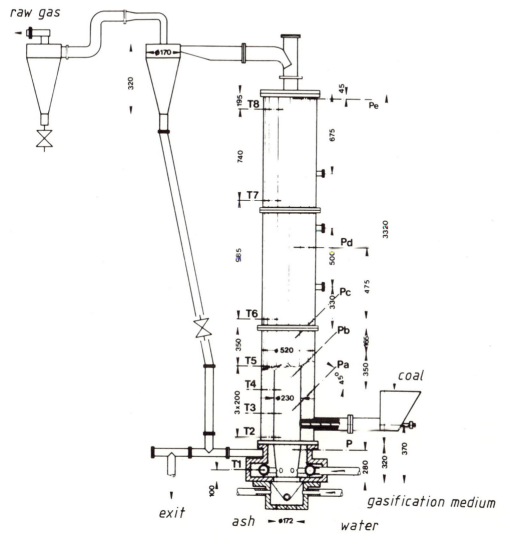

Fig. 34. High temperature Winkler
fluid. bed gasifier (14)
(Rheinbraun)

Entrained bed with transporting stream of gasification-
and combustion-gas; coal blown-in downwards inclined in
a swamp of slag; the highest temperature in the swamp;
outlet of liquid slag at the bottom; gasification reaction
during transport in the overlying pipeline. (Fig. 35)

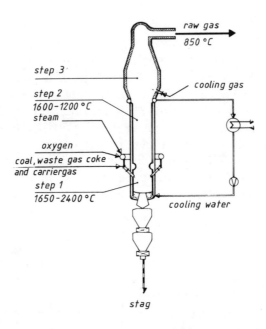

Fig. 35.   Entrained bed gasifier (15)
(Saarberg, Otto)

Fluidized bed exclusively whirled up by means of gasification; heat supply by heat exchange tubes. (Fig. 36)

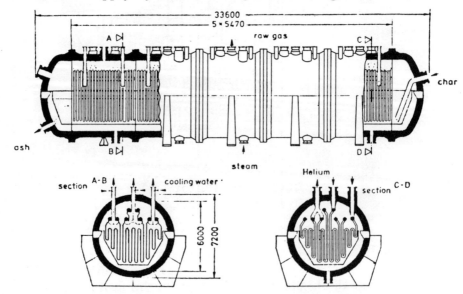

Fig. 36.   Fluidized bed gasifier for nuclear heat (12)
(Bergbau-Forschung)

Production costs

Figure 37 summarizes an estimation of SNG- and H_2-
production costs. Naturally SNG-cost is more dependent
on the cost of coal in the case of conventional gasification. For
today's hard coal costs in Western Europa, of about 20 DM/Gcal,
a reduction in the gas costs of 30 % can be achieved. Moreover,
the gas costs are in the case of nuclear gasification less depen-
ded on the costs of coal. Comparing SNG and H_2 it is expected,
that the production of H_2 leads to lower prices. This is mainly
due to the higher efficiency and the simpler gas processing in
the case of H_2 production. (Fig. 37)

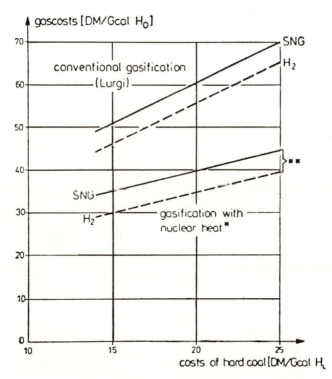

** credit of electric power 5 Dpf/kWh

streamfactor 8000 h/a

Fig. 37. Gas production costs (2)

COAL LIQUEFACTION

Coal contains liquid hydrocarbons. We set them free only by heat effects, just as gaes and solid distillation residue with pitch. Such a process is called "pyrolysis". Liquid hydrocarbons make up only about 10 % of coal. That is too little for attaining liquids from coal in a quantity capable of substituting petroleum.

Coal and oil

Petroleum is easy to handle in consuming and refining much more so than coal. Coal really could never regain the petroleum market, if not for the view, that petroleum reserves will run short at the end of this century at this rate of consumption.

Figure 38 shows the expected mined quantities of coal and petroleum in the world. Because of the inequality of consumption and surces coal will have support petroleum until coal itself runs short, too.

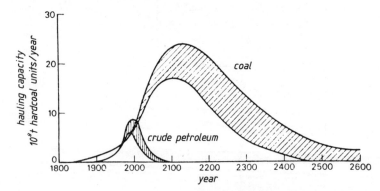

Fig. 38.   Expected world production of
coal and petroleum   (4)

Today 0,3 million tonnes of liquid and gaseous primary products per year are made by gasification and synthesis at SASOL in South-Africa. 60 - 70 % of these products are motor fuels. From a new plant - SASOL II - we expect 2,1 million tonnes per annum. Coal mining is comperatively easy in South-Africa.

During the second world war Germany was cut-off from

petroleum-sources. Only 13 % of motor fuel was derived from
handy petroleum (Fig. 39). In this energy situation 12 hydrogenation factories converted hard coal, brown coal or pitch as
eligible raw materials to motorfuel in 1943. Hydrogenation
yielded 2,9 million tonnes and 0,35 million tonnes were produced by synthesis.

HYDRATION	2,918 MIO т	
SYNTHESIS	0,353 MIO т	
DESTIILATION OF		
TAR FROM COKE OVEN	0,128 MIO т	87 %
BENZOL FROM C.O.	0,355 MIO т	
PETROLEUM	0,583 MIO т	13 %

Fig. 39.   Motor fuel production in Germany
in the world wor year 1943   (4)

Fundamentals of Coal liquefaction

     If we aim to produce hydrocarboncompounds from coal with
its main constituent carbon we just have to add hydrogen to coal.
There are tow ways of converting the solid material coal to the
liquid material oil:

1. The direct way: decomposition of large coal molecules
    by adding gaseous hydrogen or a solvent composed of
    hydrogen. Coal-oil is produced and finally gasoline
    with a low-molecular weight.

2. The indirect way: coal conversion with steam and
    oxygen into a synthesis gas, that consists of carbon-
    monoxide and hydrogen; followed by synthesis to
    gasoline by applying pressure with a catalyst.

Influences on the composition of hydrogenation products are to
be gathered from figures 40, 41 and 42. The contents of
higher boiling liquids like raw benzine, middle oil and gaseous hydrocarbons increase with the temperature, while that
of heavy oil and sediment decrease.

A high hydrogen consumption, of course, leads to high contents of low boiling hydrocarbons having a high hydrogen concentration. The quantities of benzine and oil increase with the consumption of hydrogen, coal and extracts are deposited in lower quantities.

During the synthesis of the gasification gas reacts mainly hydrogen with carbonmonoxide to $-CH_2-$chains. Steam and carbondioxides obstruct the synthesis severely.

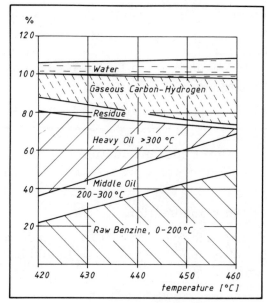

Fig. 40. Product distribution from coal
hydrogenation in swamp-phasis (17)

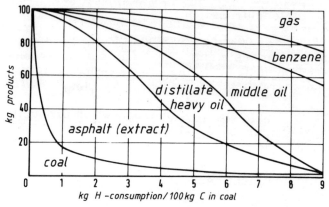

Fig. 41. Product distribution dep. of H_2-
content in coal hydrogenation (18)

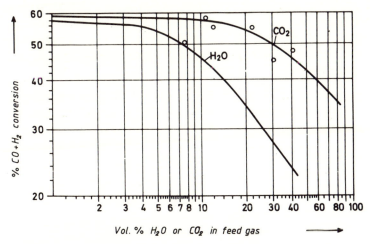

Fig. 42.   Influence of H_2O and CO_2- and H_2-
          reaction with Fe-catalyst in gas
          synthesis (5)

Processing principles

Tow ways of converting coal to liquid coal substances
have already been mentioned: hydrogenation and synthesis.
Furthermore, hydrogenation is divided into two methods.
The methods are named according to the inventors and
scientists, living in the twenties:

1. Pott-Broche. Extraction of coal by a solvent composed of
   hydrogen at temperatures of about $300\,^{\circ}C$ to $500\,^{\circ}C$
   and pressures between 10 and 500 bar. (Fig. 43 left)

2. Bergins-Pier. Treatment of coal (usually in oil suspension) with molecular hydrogen in the presence of
   catalysts at temperatures of around $150\,^{\circ}C$ and pressures of up to 700 bar. (Fig. 43 middle)

A variant of the second-metioned process is coal to let un-
dergoe pyrolysis and to hydrogenate the liberated volatile
matter. This process resembles the hydrogenate processing of high boiling sediments from raw oil distillation or
raw oil cracking processes. (Fig. 43 right)

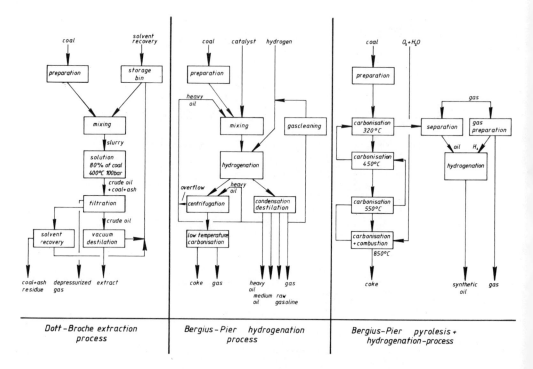

Fig. 43. Hydrogenation process

The three variants of hydrogenation are followed by five variants of gas synthesis in Fig. 44. Common to all processes is that exothermic reaction heat from transfering carbonmonoxide and hydrogen to (- CH -) chains must be quickly eliminated to keep the temperature constant at about 250 °C. The pressure rests according to the properties of the catalysts between 10 and 50 bar.

In all illustrated processes the gas must have a good and equal contact with the catalyst. This happens in the first and second cases in a fixed bed, in the third in a fluidized bed, in the fourth in an extrained bed and in the fifth case in a liquid oil phase. The cooling variants are:

1. indirect cooling by water with formation of steam
2. cooling a partial flow of circulating gas and feeding between the different catalyst beds
3. installment of cooling tubes in a fluidized bed

4. installment of cooling tubes in entrained bed
5. installment of cooling tubes in a whirled up suspension.

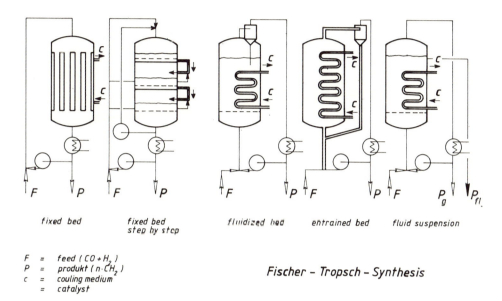

| fixed bed | fixed bed step by step | fluidized bed | entrained bed | fluid suspension |

F = feed (CO + H₂)
P = produkt (n·CH₂)
c = couling medium
  = catalyst

Fischer - Tropsch - Synthesis

Fig. 44.   Gas synthesis process

Developing projects

There are a large number of developing projects of coal
hydrogenating in Europe and overseas. There are plants in
Germany, Great-Britain. Poland just as in the United-States
of America and Japan. Most of them, however, are experimental plants in kilogramm-scale. If we selected from all the
working or constructed plants only those, whose size corresponds to the extent of the industrial plants in the case of scale
up factor 10, there would only remain three projects. Fig. 45.
One of them, the Exonprocess works according the extraction-
principle. The two others (Gulf and Ruhr-coal) run according
to the hydrogenation principle. A combination of pyrolysis and
hydrogenation was formely practiced by FMS in Princetown
(USA) with an output of 36 tons per day. (Fig. 45)

	Beginning	Process Engin.	Press. bar	Temp. °C	input t/d	output t/d	cost
Ruhrkohle Bottrop FRG	1979 (1974)	Hydrogenation with catalyst	300	475	200 (0,5)	40 gas 30 benzine 70 middle oil	300 Mio DM
Gulf Tacoma USA	1980 (1971)	Hydrogenation without catalyst	140	455	6000 (50)	420 gas 440 naphta 1945 oil (SRC II)	1 Mrd Dollar
Exon Baytown USA	1978 (1971)	Extraction	100 - 150	450	250 (1)	14 gas 32 benzine 17 middle oil 18 heavy oil	300 Mio Dollar

Fig. 45.   Coal liquifaction processes

A very clear description of hydrogenation plants is reproduced in Fig. 46 that of Bergbau-Forschung. It works on a
similar principle to the Gulf-plant. At the beginning a paste of
coal, drawn back oil and a catalysts (Fe_2O_3 - Ferric-oxide)
with almost 50 % solid material concentration is produced. The
paste is pumped by a cylinder pump over a preheater to the reactors, while circulating gas and hydrogen are added by two other
pumps. A temperature of 475 °C and a pressure of 300 bar in
the reactors is achieved by the pumping and heating process.
After the hydrogenation is carried out, a mud is separated in
a hot separator (temp. higher than 400 °C), composed solid
material and heavy oil. Light and middle oil are abstracted
from the cold separator. Other products from the separators,
liberated in consequence of pressure reduction are process
gases, just like the gases at the top of separator, which are
under pressure. The latter gases are delivered back. (Fig. 46)

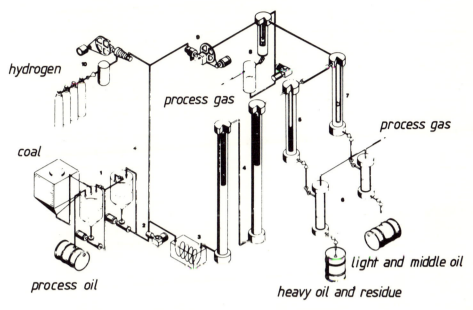

hydrogen

process gas

coal

process gas

process oil

heavy oil and residue

light and middle oil

Fig. 46.   Hydrogenation plant "Kohle - Oel"
of Bergbau-Forschung (25)

Figure 47 illustrates the extraction process by the Exon-flowscheme. The solvent is here middle oil from the process enriched with hydrogen.

It is remarkable, that as in the new Gulf-hydrogenating process, vacuum distillation is applied to separate solid deposits. Old hydrogenation and extraction plants worked with centrifuges and filters. There it was difficult to clean the tough and dirty materials. For reconditioning of the solid deposits there are two alternatives:

Carbonization to coke, gas and volatile products or gasification with oxygen and steam, e. g. in entrained-bed gasifier (TEXACO). (Fig. 47)

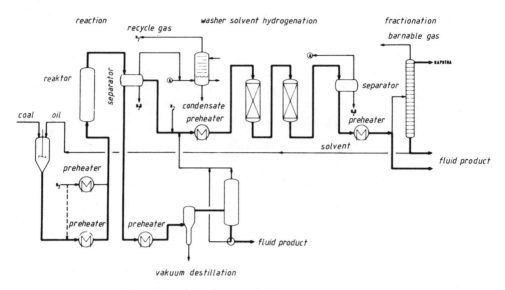

Fig. 47.   1 t/d plant of Exon (18)

The combination of pyrolysis and hydrogenation is described in Fig. 48 in the example of the COED-process (Char
Oil Energy Development). Coal and gas are in counter-flow;
the oxygen for heatgeneration is not added until the most volatile matters are expelled in order to avoid destruction of these
by combustion. The degasification and hydrogasification product
is devided into gas and oil, the gas serving hydrogenation
after its recondotioning. According to information from the operators, we obtain the following quantities from young coal:
25 % oil, 15 % gas and 60 % coke. Compared to pure pyrolysis,
the output of oil is no more than insignificantly higher and
we get lower amounts of gas. (Fig. 48)

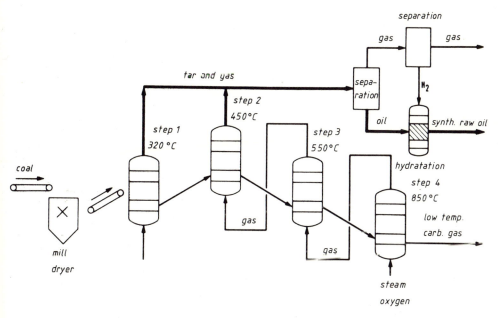

Fig. 48. COED process of FMC (18)

The other possibility of converting solid coal to liquid coal-oil or bezine is the indirect way via the gas phase. Until now this method has been used at SASOL in South-Africa, where coal is cheep: it lies in seams of 3 meter thickness and is handled in open pit mining by an inclined elevator. One-quarter of this coal is fine-coal and is supplied to power-stations. Three quarters are lumpy-coal, suitable for gasification in the Lurgi-pressure gasifier.

The SASOL-plant, built in 1955/56, is the first and up to date largest liquefactionplant with a 0,27 million tons per annum charge coal output. At the end of this year SASOL II with 2 million tons per day begins to work. A similar futher plant of the same size is in the planning stage. (Fig. 49)

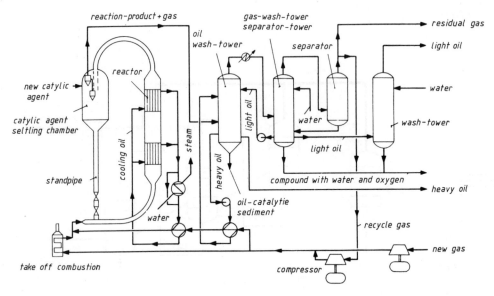

Fig. 49. Synthol process (8)

The most recent modification of synthesis at SASOL, the
Syntol-process, is illustrated in Fig. 49. It's an entrained
bed process, operating with circuled catalysts. The residence
time of gas in the extended reaction room (extension from 1 to -
meter diameter) is almost one minute. This process running
with 20 - 23 bar and 300 - 340 ^{o}C achieves a three-fold
conversion in one unit compared with one unit according to the
Arge-principle. The separating-process for liquid and gaseous
products has a similar design in both SASOL - plants.

Costs

The prices of motor-fuel as well as fuel-oil for heating
purposes from coal are not competive with the same fuel
derived from petroleum. Today the petroleum prices and
with it the prices of refined oil are high, sometimes because
of political reasons. And at the moment they are resting a
good deal higher than in the last ten years. Because of its
scarcity prices will probably be even higher in fifty years.

Mining costs of coal are included in the costs of coal refining products. They are very different according the position
of coal in the soil. We have an open pit mine and shafts about

1000 m deep. The prices of refining plants for hydrogenation lie around some thousand million DM. According a study in 1974 there was a price comparsion as reproduced in Fig. 50, with coal-charge -costs between 0,8 and 6,4 Pfennig per 10^3 kJ, that of benzine between 20 and 50 Pfennig per liter. We see that the low benzine prices in Souther-Africa correspond to the costs there. In West-Germany benzine prices from synthesis-plants would be as twice high as those from petroleum. In the meantime benzine prices in West-Germany have incrased to double the value. An approximation is, however, not fully achieved, because the plant costs are also obviously increasing.(Fig. 50)

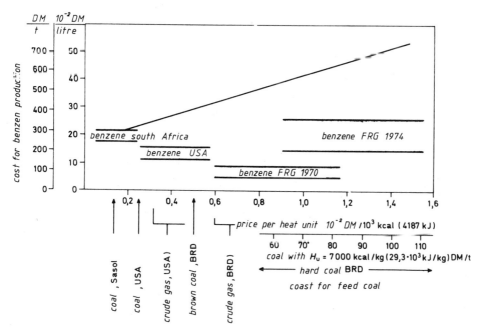

Fig. 50. Benzine production costs of Synthol process
depending on costs for the feed coal (5)

FLUIDIZED BED COMBUSTION

Combustion in fixed bed, entrained bed and fluidized bed

Until fifty years ago coal fired boiler plants worked by gratefiring (Fig. 51 on the left). Coal in pieces wanders in a moving grate over the bottom of the boiler. The combustion air blows through the layer, and the hot gases deliver their heat to the tubes, flown through by water or steam. The

ash is thrown away at the end of the wandering grate. Requirement for a good burn-out are a constant density of coal filling
and a constant distribution of combustion-air./ This was succeeded in time by coal dust firing, a technology still applied
today, illustrated in Fig. 51 on the right. On the one hand with
this process the fact was considered that most fine grained coal
with grat specific surface is the best condition for a quick reaction with oxygen and for the heat liberated thereby. On the
other hand this technology obeyed the emergency, created by
new ming-processes. Milling and slicing machines produced
hard coal of very high fineness, and even soft brown coal always broke to very fine granultes. A relatively good efficiency
was achieved by blowing fine coal with pre-heated air into the
boiler room. At very high temperatures we reached an ash
with drawing in the form of liquid, whereby cleverness in ope-
rating-technology is needed to avoid coverage of boiler-tubes
with liquid slag.

A new method of combustions, which we are still testing,
is that in the fluidized bed, as illustrated by Fig. 51 in the
middle. The fine coal is blown by combustion air through blow-
ground so strongly, that all particles are in a state of suspension. The steam tubes are immersed into the appanrently
bubbling hot layer; further tubes are situated in the gasroom
as in all other power processes.

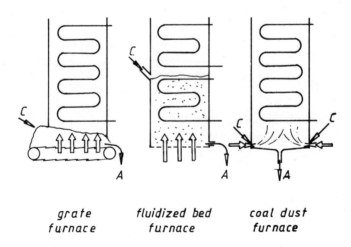

grate   fluidized bed   coal dust
furnace   furnace   furnace

Fig. 51. Different methods of coal firing

Principle of fluidized bed combustion and its advantages

Some remarks on the principle of fluidization may serve to facilitate the understanding of the behavior and properties of fluidized bed combustors. If a gas is passed upward through a bed of fine particles, the pressure drop across the bed will initially rise as the gas velocity is increased. As the pressure drop equals the weight of particles, the bed expands and the particles will be suspended by the flow. The pressure drop has reached a steady value and remains constant even at further increas of the gas velocity. This flow regime is referred to as fluidized region. If the gas velocity is increased beyond this region, dilute phase pneumatic conveying occurs. The pressure drop/velocity relationship is depicted in Fig. 52.

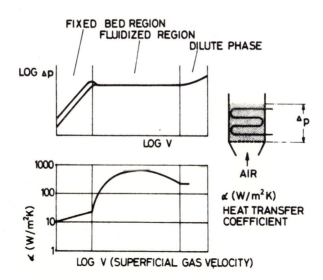

Fig. 52. Pressure drop and heat transfer in fluidized bed (19)

The fluidized bed resembles, in many aspects, the behavior of a liquid. The very intensive mixing of the particles has a favourable effect on mass and heat transfer. This results in comparatively high heat transfer coefficients, also shown in Fig. 52. The curve has a fairly broad maximum at values much higher than obtainable by convective heat transfer. This can be used to benefit by the design of immersing heat exchangers.

The main advantages of a fluidized bed combustion are summarized in Fig. 53.

■ HIGH HEAT TRANSFER COEFFICIENT
  → less tube surface

■ HIGH VOLUMETRIC HEAT RELEASE RATE
  → reduced combustor size
  lower power station capital costs

■ FUEL DESULFURIZATION IN THE COMBUSTOR
  → reduced SO_2-emission by means of
  adding limestone to the coal feed

■ LOW OPERATING TEMPERATURE
  → reduced NO_x-emission

■ LESS RESTRICTIONS ON FUEL QUALITY
  → poorer grades of coal can readily be burnt

Fig. 53. Advantages of fluidized bed combustion
          (19),

1. The already mentioned high heat transfer coefficients in fluidized beds result from the intensive gas-solids contact and lead to substantial savings of heat transfer surface.

2. The high volumetric heat release rate in fluidized bed combustors due to the excellent heat transfer conditions enables to cool the combustion zone by closely packed immersed heat exchanger banks and to control the combustion near stoichiometric.

3. In fluidized bed combustion, fuel desulfurization can be achieved directly within the combustion zone simply by adding comparatively small amounts of limestone to the coal feed. The effective gas-solids contact, together with the high affinity of SO_2 to CaO, provides excellent desulfurization. By absorption of SO_2 the limestone is converted to gypsum, which can be

readily disposed of without detrimental effects on the environment. The effect of desulfurization in the combustor by adding limestone to the coal feed is illustrated in Fig. 54.

4. The low NO_x-emission for fluidized bed combustors is due to the low combustion temperatures from 800 to 900 $^{\circ}$C.

5. The carbon concentration in the fluidized bed material is in the order of 1 percent only. Thus even high ashed coals can be burnt with sufficiently long residence times without any problems.

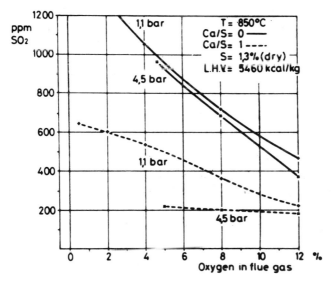

Fig. 54. SO_2-Emissions as a function of excess air, pressure and Ca/S ratio (19)

Fluidized-bed-firing-plants for atmopheric pressure

Fig. 55 and 56 illustrate two fluidized-bed firing-plants for normal pressure. In the first with a fluidized bed area of 5 m^2 a coal charge of 1 ton fine coal mud per hour is achieved. In the second case the steam production of a power plant amounts to 9 ton p.h. on an aera of 30 m^2. Reference must be paid to the different manners of coal feeding: blowing in or throwing in. In both cases a constant distribution of fuel is important. Moreover the blowing in of secondary air deserves attention, as you can see in Fig. 55.

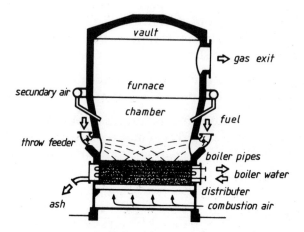

Fig. 55. Fluidized bed combustion
plant "Gneisenau" for
mineral-coal mixtures (20)

So it is possible to feed additional air for combustion in a
turbulent fluidized bed, without stirring up the layer. The
exchange of heat from the fluidized bed surface to the layer
is excelent when secondary air is added.

Such fluidized-beds are experimentally applied for the
combustion of mineral-coal-mixtures from flotation or
muds out from coal water separations. Both materials
contain a high concentration of ash and water.

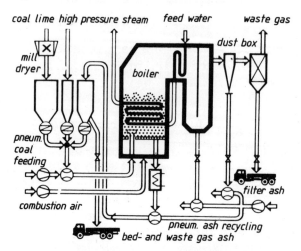

Fig. 56. Flowsheet of 6 MW-Fluidized-
Bed-Combustion "König Ludwig" (21)

Fluidized-bed-firing-plants for elevated pressures

It has already been mentioned that one advantage of fluidized-bed-firing is the grat heat transfer leading to a reduction
of heating areas and -rooms. The boilerroom of a dust-firing-
plant is as gigantic as a multi-storey building, while that of a
fluidized bed is relatively small. The steambioler in Fig. 57
has a diameter of 5 m and a height of 12 m, its thermal output
is 150 to 200 MW. Such a boiler is able to work under a pressure of 10 bar and more. Outer and inner coats are arranged
in such a manner, that the necessary combustion air flows between both walls, cooling these on the one hand and being heated at the same time, on the other. A pressure firing is not
only carried out, because the toxic emission of sulfurioxide
is becoming less, as shown in Fig. 54. We intend rather to
drive turbines with high pressure eshaust air similar to the
driving by high pressure steam. As shown in Fig. 58, the gas-
turbine delivers its effiency twice: to the aircompressor and
to an electrical generator. The remaining heat of exhausted
air serves air preheating.

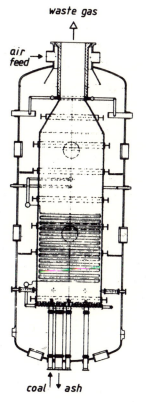

waste gas

air feed

coal　ash

Fig. 57.　High pressure boiler with
fluidized bed combustion
(22)

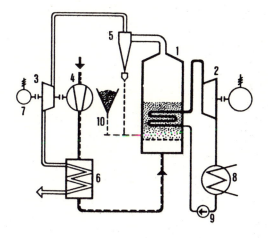

Fig. 58.　Gas-steam-turbine-plant
on basis fluidized bed
combustion (23)

The efficiency of the combined plant with electricity production on the steam and gas side, however, lies at 39,5 % and therewith higher than all comparable plants. (Fig. 59)

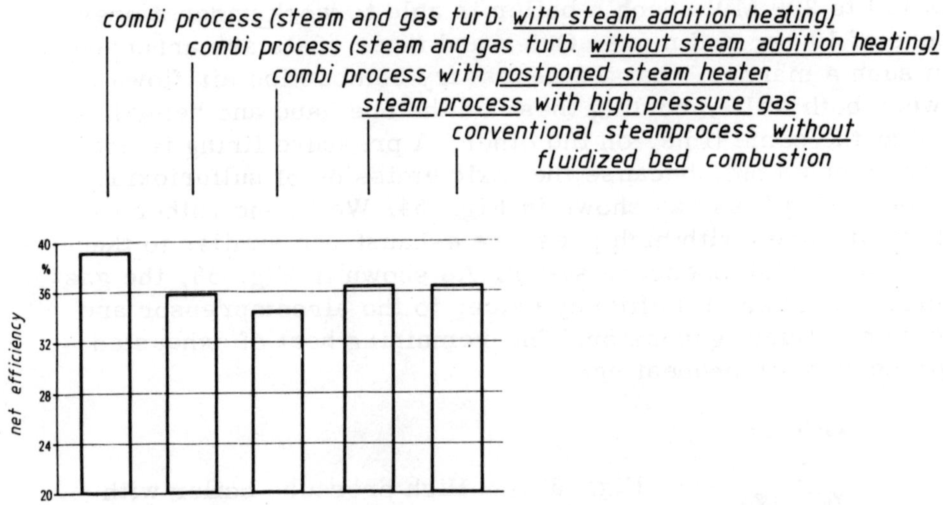

Fig. 59. Net efficiencies of power stations with high pressure fluidized bed (22)

Fig. 60 shows a comparison of different firing-plants in electricity production. It is easy to recognize, that the conventional plant (A) is much bigger, than the fluidized-bed-firing plant operating at 16 bar with the same capacity. Fluidized-bed-firing plant for normal pressure is shown between both.

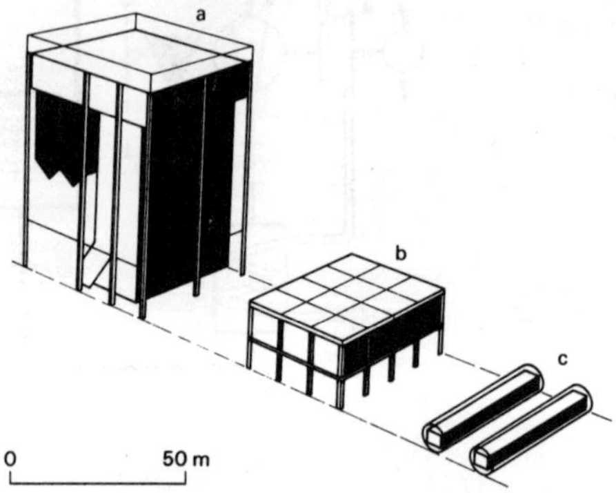

Fig. 60. Plant largeness of power stations (24)

Choice of Literature

1. Rheinische Braunkohlenwerke AG Köln (Hrsg), Braun-
                            kohlenveredelung, Verlag Die Braun-
                            kohle 1976

2. G. Bischoff ,   Die Energievorräte der Erde,
                        Glückauf 110 1974

3. Ruhrkohlen-Handbuch
                        Verlag Glückauf GmbH Essen 1969

4. G. Kölling,   Herstellung von Kraftstoffen aus Kohle,
                        HDT 405, Vulkan Verlag Essen 1978

5. J. Falbe,   (Hrsg), Chemierohstoffe aus Kohle,
                        Georg Thieme Verlag 1977

6. Mackowsky/ Wolff, Untersuchung des Koksbildungsver-
                        mögens Erdöl aus Kohle 8 / 1965

7. B. Bock,   Zur Heißbrikettierung von Wirbelschicht-
                        koks und Kohle, Dissertation Aachen
                        1970

8. F. Benthaus ,   u. a. Rohstoff Kohle,
                        Verlag Chemie Weinheim-New-York
                        1978

9. H. Jüntgen,   Die künftige Rolle der Vergasung und
                        Verflüssigung von Kohle für die Ener-
                        gieversorgung,
                        HDT 405, Vulkan Verlag Essen 1978

10. E. Ahland,   Stand der Formkoksentwicklung,
                        Fachberichte Hüttenpraxis Metall-
                        verarbeitung 1978

11. K. H. van Heek,   Stand der Kohlevergasung,
                        HDT 405 Vulkan Verlag Essen 1978

12. K. H. van Heek,   Gasification of Coal by nuclear Energy
                        Simposio Internacinal Barcelona 1978

13. G. Röbke,   Weiterentwicklung des Lurgi-Druckver-
                        gasungsverfahrens, HDT 405, Vulkan
                        Verlag Essen 1978

14. F. H. Franke ,   Vergasung von Braunkohle zu Synthese-
                        und Reduktionsgas, HDT 405, Vulkan
                        Verlag Essen 1978

15. M. Rossbach,   u. a., Druckvergasung nach dem Saarberg-
                     Otto-Verfahren, HDT 405, Vulkan Verlag
                     Essen 1978

16. Van Heek/   Materials, Problems and Research in
    Wanzl,   German Cola Conversion Projekts,
                     Erdöl und Erdgas Bd. 32 1979

17. K. H. van Heek, Coal as Raw Material and Energy Source
                     Int. School of Energetics Nov. 1975

18. J. Romey,   Stand der Kohlehydrierung,
                     HDT 405, Vulkan Verlag Essen 1978

19. H. D. Schilling, u. a., A new Concept for the Development
                     of Coal Brning Gas Turbines,
                     Gas Turbine Conference London April 1978

20. V. Asche,   Beseitigung von Flotationsbergen durch
                     Wirbelschichtverbrennung - Projekt
                     Gneisenau VDI 322 VDI Verlag 1978

21. K. G. Stroppel/J. Langhoff, Demonstrationsanlagen König
                     Ludwig und Flingern der Ruhrkohle AG
                     Essen, VDI 322 VDI Verlag 1978

22. E. Wittchow   u. a., Konzeption von Gas/Dampfturbinen-
                     anlagen mit druckbetriebener Wirbelschicht-
                     feuerung VDI 322 VDI Verlag 1978

23. H. D. Schilling, Die Wirbelschichtfeuerung - Einsatz-
                     möglichkeiten für die Strom- und Wärme-
                     erzeugung aus Kohle
                     VDI 322 VDI Verlag 1978

24. H. D. Schilling, Die Wirbelschichtfeuerung als neue Tech-
                     nologie zur Strom- und Wärmeerzeugung
                     aus Kohle
                     Glückauf Nr. 3 1978

25. B. Strobel   u. a., Ergebnisse der Versuchsanlage
                     KOHLEOEL der Bergbau-Forschung,
                     HDT 405, Vulkan Verlag Essen 1978

LIST OF LECTURERS

A. Blandino
Tecnomare S.P.A.
Venice, Italy

Burkhard Bock
University of Essen
Universitatstrasse 3
43 Essen 1, Germany

Robert Budnitz
Office of Nuclear Regulatory
  Research
Nuclear Regulatory Commission
Washington, DC 20555

Henry Ehrenreich
Division of Applied Sciences
Harvard University
Cambridge, MA 02138

Walter Marshall
UKAEA
11 Charles II Street
London, SW1Y 4QP, U.K.

Luigi Paris
ENEL
Via G.B. Martini, 3
Rome, Italy

Daniel Sewer
Scientific and Technological
  Affairs
United States Embassy
Via Veneto
Rome, Italy

Melvin Simmons
SERI
1536 Cole Boulevard
Golden, CO 80401

Chauncey Starr
EPRI
3412 Hillview Avenue
P.O. Box 10412
Palo Alto, CA 94303

Richard Wilson
Lyman Laboratory of Physics
Harvard University
Cambridge, MA 02138

LIST OF PARTICIPANTS

Fernando Amman
Stig Arff-Pedersen
Francisco Arjona
Carlo Maris Bartoline
Friedhelm Bosten
Gabriele Botta
M. Bougerra
Alys Carrus
Stefano Chow
Pierdomenico Clerici
L. Myrddin Davies
Pierluigi Gradari
Alfanso Graziaplena
Guiseppe Guarino

Peter Laut
Guglielmo Liberati
Vicenzo Nassisi
Rafael Negrillo
Michel Oria
Giorgio Pagliarini
Mirella Ricci Ghigo
Giancarlo Sacerdoti
Thomas Sanford
M. Isabel R. Soares
Oliverio D.D. Soares
Eugenio Tabet
Jose Vara
Roberto Vigotti
Raffaele Visco

1. R. Negrillo	13. M. Ricci Ghigo	25. D. Serwer
2. H. Ehrenreich	14. G. Botta	26. A. Graziaplena
3. L. Paris	15. P. Laut	27. R. Wilson
4. R. Budnitz	16. L.M. Davies	28. F. Bosten
5. R. Amman	17. R. Vigotti	29. A. Carrus
6. M. Bouguerra	18. V. Nassisi	30. C.M. Bartolini
7. C. Starr	19. G. Pagliarini	31. P. Gradari
8. G. Guarino	20. B. Bock	32. S. Chow
9. T. Sanford	21. I.R. Soares	33. G. Liberati
10. S. Arff-Pedersen	22. W. Marshall	34. Unknown
11. W. Jentschke	23. O. Soares	
12. P. Clerici	24. E. Tabet	

INDEX